*Chart of Cook Strait, featuring areas that became part of Richard Barrett's life. Produced by James Wyld, London, 1771.*

ALEXANDER TURNBULL LIBRARY

# The Interpreter

## Tribute to Mr Barrett

Whooh! Them er stirring times, when Chief Dickey did reign
O'er Moturoa's realm on't flat lung the shore,
To bir such a mind, be'ent us owd whelers faim,
and close by't firehob the yirns tow spin o'er.
The saucy Whykatos, gowd Lard! didn't um squelch!
As um Pukerangioro come tow repeat:
When oot fram't tunnels, didn't our guns let belch!
An in scyores tuk the beggars clin off of thir feet!
Bravo, Dickey Barrett! ul roun shout Bravo!
Wae such Chief ne'er an airmy nid turn tail on foe.
Bravo, Dickey Barrett! ul roun shout Bravo!
Ner er may owd England of such stuff run low.

Hood, A. (1890) *Dicky Barrett with his Ancient Mariners and
Much More Ancient Cannon at the Siege of Moturoa:
Being a Realistic Story of the Rough Old Times in New Zealand
among the Turbulent Maoris and the Adventurous Whalers ere Settlement Took Place.*
TARANAKI HERALD 1892

# The Interpreter
## The Biography of
## Richard 'Dicky' Barrett

~

### Angela Caughey

**David Bateman**

*Dedicated to all Dicky Barrett's descendants*

First published in 1998 by David Bateman Ltd, 30 Tarndale Grove, Albany, Auckland, New Zealand

Copyright © Angela Caughey, 1998
Copyright © David Bateman Ltd, 1998

ISBN 1 86953 346 1

This book is copyright. Except for the purpose of fair review, no part may be stored or transmitted in any form or by any means, electronic or mechanical, including recording or storage in any information retrieval systems, without permission in writing from the publishers. No reproduction may be made, whether by photocopying or by any other means, unless a licence has been obtained from the publisher or its agent.

Cover design by Sue Reidy
Book design by Maria Jungowska/Scope
Printed in Hong Kong by Colorcraft Ltd

## CONTENTS

|   |   | Page |
|---|---|---|
|   | Acknowledgements | 6 |
|   | Prologue | 9 |
| 1 | *March 1828–February 1832* ~ Trading at Ngamotu | 21 |
| 2 | *April 1832–June 1833* ~ Siege, Migration and Hardship | 41 |
| 3 | *June 1833–27 August 1839* ~ Whaling at Te Awaiti | 61 |
| 4 | *28 August–25 September 1839* ~ Buying Wellington | 78 |
| 5 | *25 September–4 October 1839* ~ Champagne, Gun Salutes & Opposition | 100 |
| 6 | *4 October–28 November 1839* ~ Middleman in the Sounds with Te Ati Awa and Ngati Toa | 113 |
| 7 | *28 November 1839–21 February 1840* ~ Buying Taranaki as Sole Agent | 125 |
| 8 | *21 February–18 April 1840* ~ The Colonel's Right-hand Man | 138 |
| 9 | *18 April–12 December 1840* ~ A Popular Hotelier | 153 |
| 10 | *12 December 1840–23 February 1841* ~ Siting New Plymouth | 171 |
| 11 | *23 February–11 August 1841* ~ A Whaler Again in Hungry New Plymouth | 185 |
| 12 | *11 August–31 December 1841* ~ Debts, Destitution and Land Disputes | 200 |
| 13 | *13 January 1842–27 August 1843* ~ No Longer Centre-stage and the Company Falters | 214 |
| 14 | *27 August 1843–23 February 1847* ~ On the Stand and Last Days | 235 |
|   | Glossary | 258 |
|   | Sources and Notes | 263 |
|   | Bibliography | 280 |
|   | Index | 291 |

## ACKNOWLEDGEMENTS

A work of this type is not written without significant input from a great variety of people. I have kept ever-lengthening lists of the dozens of people who have contributed but, distressingly, the names and phone numbers of some special individuals, mainly Barrett connections, have been inadvertently lost. These were written into a small notebook that sat on a shelf beside my computer — and just above my rubbish tin. It must have fallen off a few months ago while I was pulling out the main phone directory. If these people read this, and do not appear in the next paragraphs, please understand this regrettable mishap and accept my sincere thanks for so willingly supplying pieces of the Barrett jigsaw or helping me to fit it together.

Some of the people I want to thank may have completely forgotten me, such as the man unknown to me who had been involved in reconstructing an ancient waka. He had been mentioned as an authority and responded instantly to my unexpected phone call. Knowing the usual paddling speed of waka, he worked out on the spot the time that Te Puni and Te Wharepouri's canoes would have taken — in 1839 — to reach the *Tory* from Ngauranga or Petone, once those two chiefs saw the ship rounding the heads.

The staff at the following libraries gave me invaluable assistance: Alexander Turnbull (Wellington), Hocken (Dunedin), New Plymouth, Wellington, St John's College, Kinder (Auckland), Auckland Museum and the University of Auckland. It is always comforting to know that the specialists in manuscripts, books, pictures, maps and newspapers are available with informed knowledge and unselfish effort for the valid enquirer and, as always, they and the librarians at Auckland Public Library's New Zealand and Pacific Section (now the Family Research Centre) and Special Collections, and my local Remuera Library have been most dependable. They often work under pressure, and I am grateful for their skills. I am grateful also for the competence of the staff at Te Whare Tohu Tuhituhinga O Aotearoa–National Archives, the Otago Early Settlers' and Taranaki museums, Wellington and Auckland Maritime museums, Toitu Te Whenua — Land Information New Zealand in Auckland and New Plymouth — the Waitangi Tribunal, the Automobile Association in Auckland and the New Zealand Geographic Board. From across the Tasman I received interesting information from the National Maritime Records and the State Library of Tasmania and Society of Australian Genealogists in Sydney. I spent valuable hours well-attended anonymously in the Sydney Mitchell Library and the Tasmanian Museum and Art Gallery have provided last-minute rushed picture service. In England, I received willing help from the Maritime Information Centre at Greenwich, the Public Records Office and the National Portrait Gallery.

Of the individuals to thank, however, Ron McLean would have to come first. He wrote his master's thesis on his great-great-grandfather, Richard Barrett, and over four years gave me unstinting time and access to his unique knowledge. Regrettably, it was not possible to give more space to his research on Barrett's parents and family. Maria Jungowska, my recently appointed editor, has been a patient and warmly accessible contact over the past three months, during the long hours of work required to complete the myriad details connected with this biography. Life would have been intolerable without her at the end of my phone and fax, giving her immediate attention to my varied needs.

In the next group are all those others, whose extra efforts have made indispensable contributions to this time-consuming exercise: Craig Ashby, Angela Ballara, Bryan Bartley, Marie-Nui and Ralphe Biss, Tracey Borgfeldt, Leanne Boulton, Yvonne Burns, Sue and Sam Chapman, Mary Cobeldick, Aroha Reriti Crofts, Charles Crofts, Jane Davis, New Plymouth's Devon Hotel staff in 1995 and 1997, Mary Donald, Neil Earley, Nola Easdale, Rosemary Entwistle, Bev Fairclough, Ray Fargher,

Phil Garland, Valerie Garton, Sharon Gemmell, Ian Gunson, Mione Hanrahan, Grant Hawke, Nell Heatley, Charles Heberley, Heather Heberley, Matthew Hockney, Pat Hodgson, Bill Honeyfield, the two John Honeyfields, Kathleen Hook, Janet Horncy, John Ingram, Loris Johnson, Peter Kingston, Vivienne Leys, Bev Louis, Don Love, Jim Love, Island Love, Peter McCurdy, Hilda McDonnell, Lloyd Macky, Bruce Marler, Craig Mills, Marion Minson and her colleagues, Verna Mossong and the friendly Auckland Methodist Archives group, Wayne Mowatt, Jodie Murray, Sheila Natusch, Mildred O'Connor, June Opie, Ann Parsonson, Kathryn Patterson, the late Ashley Pettit, Kerry Phillips, Grant Phillipson, Frances Prentice, David Retter, Helen Riddiford, Jinty Rorke, Stuart Scarle, Rene Sharman, Ian Shearer, David Suisted, Linda Tancred, Mike Taylor, George and Lois Tikao, Shay Turnbull, Ray Watembach, Alexander Watene-Watson, Shirley Watson, and Marion Wellington.

And, of course, my immediate family and friends have my gratitude for their patient support and interest as the monumental task took my attention away from them, my basic, central, much-loved anchors.

~

# PROLOGUE

You might say it all began for Richard Barrett in 1826, when an audacious Englishman, Edward Gibbon Wakefield, asked his younger brother, William, to help in the abduction of a fifteen-year-old heiress. Barrett was then an unknown seaman but his name was to become noteworthy, and the British Government was to be catapulted into settling New Zealand as a result of Gibbon Wakefield's persuasive proposals for colonisation, dreamed up while he was in prison after the unsuccessful bid for the heiress. The story of the abduction and its aftermath is worth telling because it indicates the character of the man who stimulated the rush of British settlers to New Zealand, and of his brother, for whom Barrett later worked.

\* \* \*

Edward Gibbon Wakefield (whom Barrett never met) was born probably in London in 1796. Quick-thinking and adroit in argument, he could charm a chicken out of a fox's mouth. After a rumpty childhood, during which the rebellious Gibbon left every school he attended under a cloud, he eloped with a young heiress, Eliza Anne Pattle, when he was twenty; yet he managed to win over her family so totally that by the time she died having their second child four years later, they had endowed him munificently.

Gibbon Wakefield was not wealthy enough, however, to buy the seat in Parliament he coveted, nor to attain the comfortable lifestyle to which he wished to become accustomed. For this he needed another heiress. After six years' research he found fifteen-year-old Ellen Turner and asked his younger brother, William, to help him entice her out of her boarding school, from where Gibbon fled with her to Gretna Green.[1] Following the Turners' retrieval of their daughter, the affair became the topic of lurid newspaper columns and London drawing-room gossip for months, and the two brothers were sentenced to prison for three years.

They were held in different penitentiaries. William's lifestyle during his incarceration in Lancaster Castle is unrecorded, but his young wife, Emily Sidney, whom he had married only in 1826, died while he was in gaol, leaving him with a baby daughter, Emily. Gibbon was sent to Newgate, where he is known to have served his sentence in comfort bought with money from the Pattle endowment. He entertained his friends, educated his two children, Nina and Jerningham, and read widely, while sympathising with fellow convicts who were penniless and suffering abysmally, being held indefinitely for mere debt. In such circumstances, Gibbon evolved a theory of 'systematic colonisation', whereby the poor could be helped and profits made. In July 1829, while still in prison, he published his ideas in a pamphlet, 'Sketch of a Proposal for Colonising Australasia', in which he suggested forming a company for that purpose.

For many decades, the industrial revolution in Britain had been creating immense wealth for some and multiplying red areas of the British Empire on world maps. On the obverse side, it was creating wholesale destitution and unemployment as the jobs of skilled workers were replaced by machines. Thousands of these suddenly poverty-stricken people chose to emigrate. Some paid their own fares, others travelled at government expense in the few experimental assisted-emigration schemes or as convict labour. Virtually all suffered severe depredations, and caused social problems in distant parts of the Empire.

Gibbon Wakefield's response to this social inequality demonstrated two of his marked characteristics: his sense of social justice — inherited from his father, his grandmother and his father's cousin, Elizabeth Fry, who were all eminent social reformers — and his sharp business acumen.

He envisaged buying land very cheaply — at a 'sufficient price' — from indigenous peoples in areas where Britain had influence, then selling it on to speculators and gentlemen settlers for a much higher sum. Twenty-five per cent of the difference would go towards the cost of running the

PROLOGUE

*Edward Gibbon Wakefield, the strongly built, fair-haired, blue-eyed charmer, who designed systematic colonisation for people seeking utopia on the other side of the world. A fond, attentive father, full of ambition and self-confidence, he flattered disarmingly and ignored or twisted the truth to his own ends (Burns 1989).* ALEXANDER TURNBULL LIBRARY

enterprise and producing profits for the shareholders, and 75 per cent would be set aside to transport workers of both sexes, disadvantaged by these industrial changes, to the new country. These emigrants were to provide the labour to break in the gentlemen's lands[2] and look after their employers' everyday needs. Eventually they would be able to buy their own land, but low rates of pay would ensure they laboured for many years to save enough to do so.

Gibbon Wakefield's concept enraptured hopeful and class-conscious empire builders all over Britain. Lords, viscounts, government ministers and people of substance — many of whose names now figure on the map of New Zealand — already thought to make a profit from colonisation. They were the people in touch with reports from colonial officers, business agents and missionaries around the world, and staunchly supported Gibbon Wakefield's ideas in branches of government and through the newspapers. So when Gibbon emerged from prison, he never again became involved with an heiress. With influential backing, he turned his attentions to implementing his scheme, and Richard Barrett, among thousands of others, became involved instead.

In the late 1820s, the results became known of one early English business expedition to New Zealand, organized before Gibbon Wakefield announced his theories and three years before Barrett landed in New Zealand. The shareholders of this first New Zealand Company of 1825 hoped for large profits. They were convinced that New Zealand flax could make superior ropes for the British Navy, that New Zealand timber, the 'cowdie' (kauri) in particular, would fill British pockets with wealth, and that sealing, whaling and general trading would be lucrative gambles, as they were proving to be in many newly explored parts of the world.

They lost £20,000 sending the *Rosanna* and the *Lambton* to New Zealand under Captain James Herd and discovered their convictions were wrong. Herd, however, left their mark on the country. He named the vast anchorage in Cook Strait Lambton Harbour, in honour of the company's chairman, and 'bought' land from the inhabitants he met at Hokianga, Waiheke, Manukau and on the Waihou River, which the Europeans called 'Thames'. Then he sailed to Australia and disbanded his crew at about the same time (1827) as Dicky Barrett was joining John Agar Love in Sydney in a trans-Tasman trading venture, and the Wakefields were about to begin their prison sentences.

After Herd's abortive trip three other British companies were formed, inspired by Gibbon Wakefield's writings: an ephemeral one in 1834, the New Zealand Association of 1837 and the New Zealand Colonisation Association or Company of August 1838 — all with New Zealand as their target; but, despite receiving reports for years about the uncouth behaviour of English citizens in New Zealand, successive Colonial Secretaries consistently denied each company support. They had no official obligations there, they pointed out. It had been a myth that Captain James Cook had annexed all New Zealand. He had claimed sovereignty over only two small areas of the country. Overburdened with other troublesome colonies, they were reluctant to increase their headaches by taking over New Zealand too. In 1833, they appointed James Busby as Resident in the Bay of Islands to control the miscreants; but he was ineffective, lacking any troops or arms with which to enforce his word.

A few years later the Colonial Secretary sought other solutions and, in 1837, the Colonial Office did offer a charter to the New Zealand Association; but with the proviso that its directors accepted full financial responsibility for the settlement and running of the country. The majority of the directors backed off quickly, the charter was withdrawn and that company folded. The next year another suggestion for control of the troublesome expatriates in New Zealand found favour with the Colonial Office. It was submitted by Captain William Hobson. He had visited the country earlier on an investigating man-of-war and favoured a form of government intervention. By May 1838, having had its charter refused and almost against its better judgement, the British Government considered taking Hobson's advice and negotiating a cessation of sovereignty from New Zealand chiefs in certain areas and installing local magistrates, called 'factors'.

Four of the New Zealand Association's hard-core directors, still keeping in close touch with the powers-that-be, heard that Hobson was being groomed to be consul in New Zealand and received the impression from a newly appointed Colonial Secretary, Lord Normanby, that a pending government bill would include a clause that, in future, land in New Zealand would be available only from British Government sources. In a scramble they took the monetary gamble, formed the New Zealand Land Colonisation Company and requested the charter offered to and declined by their just-defunct organisation; but the government declined it, replying that a conditional promise made to one association did not apply to a different organisation.

The directors reacted immediately. They could not allow their long-planned dreams to be thwarted at this late stage. They twisted Normanby's words and told the shareholders that they appeared to have his ministry's support. Then they rushed ahead with arrangements to beat those government plans for New Zealand land. Starting on 21 March 1839, in a frantic six-week period they assembled £5,000's worth of goods for bartering and spent £5,250 on buying and outfitting a ship to carry them. It was one of the absurdities of the project that they considered the millions of acres of land that they hoped to acquire of less worth than the ship carrying the purchase price. They wrote instructions for the leader of the team, whom they referred to as the Principal Agent, and arranged hundreds of disparate details for the expedition.

The instructions for the team leader were wordy, complicated, and study of them reveals the ethnocentrism that permeated their attitudes; and study of the Principal Agent's later actions reveals how often he ignored, or was not able to fulfil, the directors' requirements. As for the myriad details, many of these were based on almost complete ignorance of the people that the expedition members would be dealing with and the geography of the country they were to buy. This lack of awareness led to the inordinate amount of trouble that the New Zealand Company experienced in the years ahead and turned Barrett's life on its head.

At the end of this hectic period, the directors appointed the Principal Agent — Gibbon Wakefield's brother William, at this stage an unemployed colonel with prison and a subsequent rather distinguished stint in the Spanish Legion behind him. He was to possess himself of New Zealand's soil before the British Government could act; but he was to pay the native New Zealanders only the 'sufficient price' referred to in his brother's theory, that is, as little as possible.

Gibbon Wakefield had precipitated this chaotic scramble by his charismatic presence and comments at the dinner party where the new company was being formed. '. . . [P]ossess yourselves of the Soil & you are secure,' he pronounced, 'but, if from *delay* you allow others to do it before you — they will succeed & you will fail.'[3] Gibbon declined to take an active part in the project, perhaps being tactful because of his criminal conviction, but also because he was fully occupied dabbling and speculating elsewhere.

At this point, Gibbon Wakefield quits the scene and William Wakefield

PROLOGUE

*The Wakefield brothers' trial took place on 23 March 1827. This sketch by an unknown artist shows William Wakefield, immaculately dressed as usual, on 14 May 1827, before he appeared for sentencing.* ALEXANDER TURNBULL LIBRARY

takes centre-stage for the company. Almost totally unprepared for the people with whom he would be negotiating and the country to which he was heading, William left his young daughter in England and sailed on the *Tory*, on Sunday 12 May 1839, complete with his small expeditionary team and his instructions. Late August 1839 saw them arrive in the South Island of New Zealand.

One of the first white men Wakefield met was Richard Barrett, now a whaler, living with his Ati Awa wife and family at Barrett's whaling station at Te Awaiti, on Arapaoa Island in Queen Charlotte Sound. Wakefield liked Barrett immediately; and, perhaps conscious of his own inadequacies for the task ahead, was impressed by the fact that the whaler spoke Maori and seemed to know the coastline intimately. He engaged Barrett as pilot and interpreter, and this cheerful, poorly educated, roly-poly seaman joined the group which finally forced the British Government to add New Zealand to its Empire. But Barrett was not well rehearsed for his role in the spotlight of history.

*An 1830s map of New Zealand, showing the areas of land that William Wakefield hoped to buy for the New Zealand Company, between latitudes 38° and 43° on the West Coast and 41° and 43° on the East Coast.* AUCKLAND INSTITUTE & MUSEUM

PROLOGUE

\* \* \*

Little is known about Richard, or Dicky, Barrett's background. He is reputed to have been born in about 1807, either in Durham or Bermondsey, in England. Neither area would have been salubrious, but both have the shipping atmosphere that could connect them with Barrett. Apart from its imposing cathedral and castle, and various other historic buildings, most of early nineteenth-century Durham was an industrial sprawl of dirt and poverty — no jewel in anyone's background; but it did sport ship-building yards. Bermondsey, in London, was equally unattractive — 'a place of slums and alleys', and 'a dumping ground for the dirtiest trades that had been shut out of the City'; but it gave rise to many sailors, being on the south bank of the River Thames.

As with his birth, nothing can be established about Barrett's early years. There is a possibility he may have had four brothers and two sisters[4] and a little education. His grammar and spelling, in two letters that survive, were weak, but he was not illiterate and his signature was clear and firm. It seems that neither of Barrett's possible places of birth had sufficient employment prospects to keep him at home as he grew into young manhood and that he opted for a life at sea. Family legend, contained in his probably faked journal, has him sailing for southern waters in a whaling and trading ship when he was about sixteen. Whales and trade certainly filled the rest of his life.

Richard Barrett, the known person, came to New Zealand in 1828. He set up a trading post at Ngamotu in Taranaki, Te Ika a Maui (the North Island) in partnership with his friend, John Agar Love, and chiefs of Te Ati Awa, the most powerful tribe in that region. He and Love later fought with Te Ati Awa against Waikato, Ngati Maniapoto and other tribes, migrated south with Te Ati Awa to the Kapiti Coast, then left them to set up independently as whalers in Queen Charlotte Sound. Six years later, after joining Wakefield, Barrett helped the Principal Agent buy land from Ati Awa chiefs at Whanganui-a-Tara, later to be known as Wellington, and from Ati Awa and Ngati Toa chiefs in Te Wai Pounamu (the South Island); then he, by himself, bought an immense slice of Taranaki for the New Zealand Company, while Wakefield went north to explore prospects there. At the end of this episode, Barrett's performance seemed to have produced a first-rate outcome for all concerned. The New Zealand Company had millions of thousands of acres, Wakefield had a very impressive-looking document signed by all the tribespeople present, the sellers had a satisfying selection of European goods (with more double-barrelled guns to come), and Barrett was promised £400, with which he intended to build a hotel.

Within three and a half years, these apparently excellent results were in disarray. The settlers faced unpleasant opposition from the tribespeople, and the colony's governor, FitzRoy, reduced their land areas drastically, Wakefield found himself in court trying to substantiate how forty-seven Ati Awa — some of them children — could sell tribal lands belonging to absent thousands, the vendors never got the double-barrelled guns they were promised, and Barrett had to sell his hotel to pay his debts.

Although Gibbon Wakefield and his associates in Great Britain never ceased their propaganda campaign against Her Majesty's Government as being the cause of these misfortunes, there could be several explanations for these debacles. Barrett could have been self-serving in the way he steered Wakefield towards his Ati Awa family and friends at Whanganui-a-Tara, the first place where Wakefield wanted to buy land, or this could have seemed an obvious opening move to a man in his position, thrown into the role of adviser without previous warning. Later, Barrett may have deliberately advantaged the few relations of his wife when he bartered for land at Ngamotu, or he may have been following Wakefield's instructions to the best of his knowledge — or lack of it. These are only two aspects of many to consider, while examining his activities as one of the focal players in the historical drama of early European settlement in New Zealand.

In the background, immensely influential European institutions were affecting the stage directions for the take-over of New Zealand. These factions comprised the British Government (and its other arm, the British Civil Service), four different companies already mentioned, with New Zealand as their focus and culminating in the 1838 New Zealand Land

Company, and the Anglican and Wesleyan churches. They were all run by individuals possessed with enthusiasm, ranging from measured and intelligent to fanatical and dishonest. Only one or two had ever been to the country.

Involved unknowingly, initially, in the drama, were other institutions — the powerful tribes of the country later to be known as Aotearoa: Te Ati Awa, Ngati Toa, Ngati Maniapoto, Waikato and Ngai Tahu, among many others. They formed 'a homogenous commonwealth of autonomous tribal states, whose boundaries, like those of Europe, fluctuated somewhat but were well known'.[5] Within this structure they formed treaties and alliances, and maintained a balance of power, in a lifestyle with a well-regulated and viable economy. But the word 'institution' is a neutral screen for the individuals within, and the people involved in New Zealand affairs exhibited the usual immense diversity of personality, character, morals, values and attitudes found in any community. Sometimes they acted in well-rehearsed roles. At others, they ad-libbed their parts, creating an on-going play that spun along almost out of control.

Barrett could have been used by Wakefield; or, conversely, Barrett may have tried to use the opportunities, which William Wakefield's arrival presented, to increase his own fortunes. Complications must have arisen from the differing expectations or private agenda of the people involved, and consideration of their actions and words, taken with their apparent strengths and weaknesses, inferred hopes, ambitions and disappointments, may further clarify the background to the settlement of New Zealand.

To help evaluate the importance of Barrett's role in these transactions, most known references to him may be gleaned by reading E. Jerningham Wakefield's *Adventure in New Zealand* (while being aware that this book is very intent on promoting the 1839 New Zealand Land Company), and the journals, diaries and reminiscences of William Hayward Wakefield, Frederic Alonzo Carrington, Ernst Dieffenbach and other less-well-known early New Zealand adventurers. Other glimpses are to be found in the many weighty volumes known as the *Great Britain Parliamentary Papers*. These contain the records of cross-examinations of people involved in the settlement of New Zealand, by British Parliamentarians trying to unravel the legality of the whole business.

Further questionable personal information is contained in Barrett's so-called journal. Ronald McLean, Barrett's great-great-grandson, analysed this mixture of truth and family myth in his M.A. thesis about Barrett. He concluded that the journal was probably written by Barrett's great-granddaughter's husband, W.T. Duffin, from his wife's memories, possibly as a contribution to New Zealand's centennial celebrations in 1940. It contains many mythical details of grand doings, often found in family sagas to substantiate the historical worth of the family subject; but germs of fact could occur within its pages. It is referred to in the text only occasionally to highlight a point, or to throw up a contradiction, and then usually in a note.

This, the main bulk of Barrett material, has been rounded out by clues found in assorted records of early days (listed in the Sources for each chapter and the Bibliography), pieced together to form a fairly reliable picture of Barrett's life.

Barrett's time in New Zealand is presented here in three sections. The first part covers his life from the period when he arrived in Ngamotu until the arrival of the *Tory*. The second carries him through his crucial activities, as he bartered for land on behalf of the New Zealand Company, and then lived among the settlers who poured in. The third section builds up an impression of his life in New Plymouth, after he and his family returned to Ngamotu, where he died prematurely.

The impression gained of 'Dicky' is of an amiable, respected, simple character, thrust into unexpected limelight. He appears to be a mixture of mainly attractive attributes, woefully out of place on the important stage of history he occupied by chance. Had he not left Britain, his name would have remained as unknown as the names of his family. His origins are uncertain, his life thus far almost undocumented, but there is no doubt that he is worthy of note as one of New Zealand's earliest European legends.

*Chapter 1*

MARCH 1828–FEBRUARY 1832
TRADING AT NGAMOTU

Towards the end of January 1832 several Waikato warriors walked into Ngamotu, an Ati Awa village on the coast of Taranaki. They were messengers sent by their principal chief, Te Wherowhero, to announce that Te Wherowhero's Tainui taua was advancing to attack the villagers at Ngamotu. A few weeks previously, Te Wherowhero and his force, comprising Waikato and Ngati Maniapoto warriors, had laid siege to Te Ati Awa in Pukerangiora pa on the Waitara River, had massacred their traditional rivals when they finally attempted to break out and then feasted on them. Now Te Wherowhero was following the custom of tribal warfare that required him to inform other Ati Awa, further south at Ngamotu, of his intentions now to attack *them*.

On 30 January, the heralded Tainui host appeared, heavily armed and filing along the seashore, watched by Ati Awa, who had taken refuge behind the palisades of their local fort, Otaka pa. At the Huatoki stream, the taua divided. A small number continued along the beach and took up positions in front of the pa. The main group turned inland and spread out, surrounding the defenders with 1,600–2,000 warriors, each carrying at least two European muskets and supported by many women and the large number of slaves captured at Pukerangiora.[1]

Among Ati Awa villagers staring out at the encircling Tainui party was a small group of Europeans — Richard 'Dicky' Barrett, his friend, John Agar 'Jacky' Love, and several English seamen. They had been living and trading at Ngamotu for nearly four years, as agents for two Sydney merchant partners, and all that time the threat of this attack had been in the air. Because of it, they had earlier bought three cannon and one swivel gun from a passing ship and, after hearing that an army was coming, they had had this artillery manhandled up the hill and into Otaka. It brought a small feeling of security but, although they had plenty of gunpowder, they had very little ammunition. Now Barrett and Love improvised. They organized one group of villagers into making powder bags out of any available material, and another into collecting a motley assortment of stones, pieces of iron and broken bottles — anything that could be used as deadly projectiles in makeshift shot. From being sought-after and protected English traders, they found themselves now threatened husbands and fathers in their Ati Awa tribal family.

The defenceless members of the village — the elders, the children and many of the women — had already moved to the safety of other pa on the top of local heights. A few women had chosen to remain but, besides the seamen, the defenders numbered only 250–350 Ati Awa, armed with traditional weapons and about 100 muskets, probably brought in by Love and Barrett. In England, the worst that the later two could have expected from neighbourly antagonism would have been drunken abuse or brandished fists or hayforks. In New Zealand, Barrett and Love and their friends were facing warrior neighbours from a very warlike society, intent on violent retribution for a previous defeat.

\* \* \*

The first historical records of Richard 'Dicky' Barrett place him on the small Australian ship *Adventure* as it sailed south along the shores of Te Ika a Maui (New Zealand's North Island) in mid-March, 1828. Barrett was first mate and the captain was John Agar Love. Physically the two men were complete opposites: Love was tall and fair-haired, while Barrett was brown-haired, short and stocky. The members of the crew whose names survive were: Billy Bundy, John Wright, Bosworth, William Keenan (a red-head), Daniel Sheridan, George Ashdown and 'Scipio' Lee, an African-American cook. All were on their first trading voyage to New Zealand and were destined to remain together for several years.

They had a questionable assignment. While good business had been

*Richard 'Dicky' Barrett, known to Ati Awa as Tiki Parete. The only image thought to be of Dicky Barrett, a woodcut by an unknown artist.* TARANAKI HERALD/*TARANAKI MUSEUM*

rumoured by other traders, the locals were known to be unpredictable — and cannibals, to boot, and mention of them brought vague memories of the fates of early explorers in the Pacific — du Fresne and Cook, and the crew of the *Boyd*, all of whom had been killed, and some eaten.[2] Sure enough, as the *Adventure* arrived off the Taranaki coast, two long canoes, powered by dozens of paddlers, left the shore and headed towards her, overtaking the small ship as she reached Cape Egmont. Any fears of an attack on this occasion, however, were soon put aside. Barrett is reputed to have had a smattering of the Maori language[3] and at this meeting, in probably his first effort as an interpreter in New Zealand, he quickly uncovered the intentions of the two chiefs who boarded. They were looking for a business partnership!

The two entrepreneurs were Honiana Te Puni-kokopu and Te Wharepouri, close cousins and leading chiefs of Te Ati Awa iwi. Te Puni was then aged about fifty-three. His hair was white, his face completely tattooed, and his body carried scars from years of fighting, but he had an impressive dignity as he stepped onto the vessel's deck. His companion, Te Wharepouri, with long dark hair caught in a tie on top of his head, was some fifteen years younger, lithe and more volatile. With sign language and simple communication, the two chiefs launched into a description of the attractions they could provide on shore for the people on the *Adventure*.

'You must take your ship to Nga-Motu [*sic*], where there is plenty of muka and numerous pigs,' Te Wharepouri told them.[4] Ngamotu had been visited by a European sailing ship several years before, and the chiefs realized that having white men living with them would bring iron tools, woollen blankets and, most importantly, muskets.

Love and Barrett would have been uncertain about whether to accept the chiefs' suggestion that they settle at Ngamotu. As captain and first mate, they had not expected to have to make that sort of decision. They were agents for Thomas Hyndes and Thomas Street, well-known Sydney merchants, who had built the *Adventure* in 1827 in a joint venture for use on Australian coastal trips.[5] However, a business depression developing in Australia had made Hyndes and Street look further afield, across the Tasman Sea, where missionaries, sealers and whalers had been drawn to the opportunities offering in New Zealand since the late eighteenth century. The merchants had packed the tiny *Adventure*'s holds with domestic articles for these opportunists, articles which were now also much sought after by the native New Zealanders. Then they included munitions for the latter who, they had heard, willingly produced cargoes of dressed flax, pigs and potatoes in return for guns. The flax particularly was highly sought after for naval ropes, cordage and sails.[6] Hyndes and Street hoped this enterprise would repair their fortunes when the goods were on-sold in Britain, so they instructed Love to sail to New Zealand and bring back as much as possible.

Love may have been unsure about his principals' reaction to the chiefs' proposal that they establish a trading post at Ngamotu, but he agreed to have a look, and the two waka escorted the *Adventure* to the only safe local anchorage. This was in the shelter of six or seven high,

*Te Puni (left) and Te Wharepouri were both grandsons by the same wife of Whiti-Te-Rongomai, after whom the hapu Ngati Te Whiti was named. Te Puni, the senior chief, was the eldest son of the eldest son, while Te Wharepouri, a younger fighting leader, was a son of Whiti-Te-Rongomai's fourth child, a daughter.* ALEXANDER TURNBULL LIBRARY

conical, rock islands, covered in patches of ragged vegetation and lying in a group out from an open, west-facing, surf beach. This long stretch of black sand was dominated at the western end by one of these cones, jutting out from the mainland like an inverted ear-trumpet — a soaring rock, 150 metres high known as Paritutu. The eastern end of the beach featured black, sprawling reefs.

When the seamen landed, the many people living in villages at the base of the rock welcomed them with friendly handshakes, hongi and food, in the customary Ati Awa manner. This was not the way in which the crew were usually treated by their own countrymen; and, after more, probably stumbling, discussion, the *Adventure*'s crew decided that Hyndes

*There were 2–3,000 people living between Moturoa and Te Henui stream and over 30 pa along that coastline when Love and Barrett joined Te Puni and Te Wharepouri there in business.*

ANGELA CAUGHEY, FROM LAND INFORMATION NEW ZEALAND/TOITŪ TE WHENUA & LEANNE BOULTON and Street would agree the chiefs' trading offer was too good to refuse.

Both groups of people had fallen on their feet. Barrett and Love could now stay in one spot and have the trade brought to them, rather than waste time and effort slogging uncertainly all around the New Zealand coastline searching for it; and the hapu would have muskets delivered to their door by the *Adventure*. Te Ati Awa looked for those other benefits too — clothing, tools and utensils — and would make it worthwhile for the Europeans to stay. They soon built a large pataka, naming it Patarutu. This was to become the centre of the trading post, but its exact siting is uncertain. Ngati Te Whiti hapu of Te Ati Awa iwi, to which both Te Puni and Te Wharepouri belonged, occupied the east side of the large Pukeariki pa, on the hill above Huatoki stream, and that may have been where Patarutu was erected; but the most sheltered anchorage for visiting trading ships was in the lee of the Sugar Loaf Islands (later known as the Sugarloaves), and that area, Ngamotu (or Moturoa, as the area was sometimes called), stands out as a probability. Te Puni, who also had ties with Ngati Tawhirikura hapu, lived at Pukeariki, while Te Wharepouri, the hapu's warrior leader, lived at Te Ruakotare, or Ruataku pa, one of the kainga at the base of Paritutu. The families of Wi Tako Ngatata and Poharama, other chiefs who would feature in Barrett's life, lived nearby.

The community the seamen were joining was surprisingly large, well-organized and industrious, and spread out in separate clusters. The coast had looked deserted as the *Adventure* had sailed by. It was not. Thousands of Ati Awa lived behind the bush-clad cliffs and scrub-covered dunes. Visible or not, there were fortified pa on nearly every vantage point, and thousands of well-trodden tracks on which news was carried through the whole country with remarkable speed. Along the 6 kilometres of undulating coast which became the newcomers' immediate neighbourhood were about 2,000 inhabitants and over thirty pa. The sparsely clothed, friendly tribespeople, including numerous children, lived in scattered groups of low-built reed and clay huts, usually surrounded by racks of drying fish swinging from lines like limp pennants, and hundreds of hectares of orderly, weeded gardens, watered by plentiful streams and small rivers.[7]

```
                        TE WHITI-O-RONGOMAI
                    (AFTER WHOM THE HAPU WAS NAMED)
      m. (1) RONGOUEROA                     m. (2) TARAWHAKAUKA
             │                                        │
   ANIWANIWA m. TAWHIRIKURA              ┌─────────────────────────┐
             │                    KARAKI TE RANGI      RURU TE HAKURAMA
   ┌─────────┼──────────┐                │                         │
 REREWHA-I-  TE WHITI-O-  TE NGATORO     PAKANGA       KIRIHIPU KUPAPA KAUA
  TE-RANGI   RONGOMAI II      │             │                      │
      │          │            │             │                      │
      │     TE WHAREPOURI     │         NGATATA-I-        MERERURU TE HIKANUI
   TE PUNI ═════ m. ══════ WIKITORIA Te Muri  TE-RANGI           m.
             │             WAKA ROTA          │             JOHN AGAR LOVE/
           ROKA                              WI TAKO            HAKIRAU
```

*Lines of descent in Ngati Te Whiti, highlighting central figures in this narrative.*
FROM MATERIAL SUPPLIED BY MARIE-NUI AND RALPHE BISS, RAUMATI BEACH.

In short order, other advantages materialized for the seamen. The friendliness first exhibited by the villagers continued. Huts were built for the newcomers and they were soon found female partners. They may not have been aware that this was the tribe's method of cementing their relationships and assuring their access to European goods, as well as ensuring that the white men did not impinge on existing tribal marriages. Two high-born young Ati Awa women were chosen for Love and Barrett, as befitted the captain and first mate of the partnership's flagship.

Mereruru Te Hikanui, who came to Love, was from Ngati Te Whiti hapu, the daughter of Kirihipu Kupapa Kaua and Hana te Wharetiki, and great-granddaughter of Te Whiti-o-Rongomai I, after whom the hapu was named. Te Puni and Te Wharepouri were his descendants, too. Wakaiwa, Barrett's partner, was one of three daughters[8] of Eruera Te Puke-ki-Mahurangi and Kuramai Te Ra, and belonged to both Ngati Rahiri and Ngati Maru hapu of Te Ati Awa. Through her mother, Wakaiwa was the only granddaughter of Tautara, the ariki or paramount chief of Te Ati Awa. Barrett's knowledge of the politics of whanau, hapu and iwi, into which he was now catapulted, was to assume vital importance in the years ahead. Both young women were beautiful and dignified and probably aged about seventeen or eighteen. Twenty-one-year-old Barrett thus

*Wakaiwa, or Rawinia, the high-born Te Ati Awa woman chosen to be Dicky Barrett's wife. From a woodcut originally held by the* Taranaki Herald. *Artist unknown.* TARANAKI MUSEUM

found himself launched into a large, ready-made, important Ati Awa family, in which the newcomers formed a new, small whanau.

Tribal unions were always marked by ceremony and much jollity and feasting, especially when tribal aristocracy was involved. Brides were washed and rubbed with scented oils, dressed up in kaitaka, maro and numerous ornaments, and crowned with feather headdresses; and they were usually 'given away', much as in the Western manner. Wakaiwa towered over her short and stocky husband, both in stature and in tribal standing, but she provided him with physical comforts, companionship and a new and welcome home.

Both races must have picked up each other's language to some extent, as Patarutu became the focus for all the inhabitants in the area. It seemed appropriate to rename the *Adventure*, and unofficially she became known as *Tohora*, the tribe's word for the 'right' whale.[9] And the villagers soon began calling Barrett Tiki Parete, Jacky Love became Hakirau, and all

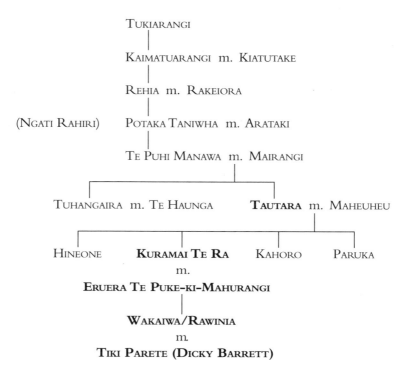

*Wakaiwa, Barrett's partner, belonged to both Ngati Rahiri and Ngati Maru hapu of Te Ati Awa. Through her mother, Wakaiwa was the only granddaughter of Tautara, the ariki or paramount chief of Te Ati Awa.*

LINES OF DESCENT SUPPLIED BY MIKE TAYLOR OF PICTON MUSEUM

the seamen's names were similarly adapted; but though the newcomers seemed to be accepted into the tribe, they were often on the outer as far as jokes or tribal practices went, some of which were quite unfathomable. This uncertain merger into Ati Awa society stretched the Englishmen's inborn attitudes and expectations. Barrett and Love, with British insouciance, may have expected themselves to become managing directors of their enterprise. Instead, they found themselves picking a wary path through the everyday affairs of which they were now a part.

As Te Ati Awa adapted, and began learning how to turn their pigs into export bacon and produce dressed flax for the British navy's ropes, the sailors inevitably had to adjust to the lifestyle and customs of the tribe.

Supplies from the *Adventure* dwindled and their eating habits changed, so birds, fish, eels, rats and dogs, cooked over fires or in hangi, now supplemented the menu.[10] They experienced novel berries, too, with aruhe, the root of the bracken fern pounded into a pottage, filling any spare spaces.

Everyone at Ngamotu seemed to be related several times over, as they sometimes were in English villages; but in England, the local gentry owned the land and the peasants and labourers worked for them, whereas in Ngamotu everyone shared not only the tribal land that they worked on, but other land, often hundreds of kilometres away, which they had claim to because of their family connections. They would reel off intricate genealogy to prove it. Every feature of their landscape had a name, and a personality and story attached. An insignificant pile of stones could mark an important tribal boundary, and everyone knew where one tribe's land finished and the next one's started, even though the areas had changed from time to time and century to century. To them it was clear. To Barrett and the other Englishmen it was confusing and seemed to have no relevance, so they listened, wondered, and put it all to the back of their minds — from where Barrett later needed to retrieve it.

Patarutu filled rapidly with goods. Ati Awa gloried in their new blankets, as the weather grew cooler, and in the novel pipes, tobacco and jew's harps, and the muskets, powder and shot which they also received for their labours. After only one month of the partnership, there were enough potatoes, bacon and pigs available to fill the *Tohora*.

While Barrett remained at Ngamotu to supervise the next load, since they took no flax on the first trip, Love sailed back to Port Jackson in Sydney, accompanied by Te Puni, Te Wharepouri, Poharama and other senior chiefs. They arrived in the first week of May, 1828, Ati Awa exulting in their first chance to inspect in person this city of unimaginable size and potential, which had spawned their new partners. In preparation, Barrett had already introduced them to barter, a practice foreign to their custom, but vital for their intentions to buy guns and ammunition for themselves.[11] The crew of the *Tohora* now understood why it was so necessary for Te Ati Awa to buy guns, and why they heard a name,

'Motunui', uttered with a special aura. Many of their new hapu friends had helped to defeat warriors from a whole confederation of Tainui tribes in the early 1820s at a place called Motunui, and had been dreading the inevitable utu ever since.

The battle at Motunui had been the result of convoluted history. Tainui tribes Waikato and Ngati Maniapoto lived next door to and inland from Kawhia, where Te Ati Awa's ancient allies, Ngati Toa, had their ancestral lands. Tainui warriors had had access to firearms since the beginning of the nineteenth century and had terrorized their neighbours with them. Ngati Toa had more than held its own, using traditional weapons in minor intertribal spats; but their principal fighting chief, Te Rauparaha, knew that Waikato and Ngati Maniapoto strongly coveted Ngati Toa's fertile coastline and could eventually conquer and annihilate them.

Te Rauparaha had therefore persuaded his people to migrate south to settle on the Kapiti Coast, and to abandon Kawhia to their inimical neighbours. In the early 1820s, he had formed a heke which, continually chased and harassed by Tainui, had travelled as far as Motunui, on its way to Kapiti, before Te Ati Awa, on whose tribal land they were, joined them to engage their pursuers. Armed only with their traditional weapons and a large amount of Te Rauparaha's guile, Te Ati Awa and Ngati Toa had achieved a somewhat unexpected victory against the Tainui fighters armed with deadly pu.

Following this, Te Rauparaha and Ngati Toa completed their trek to Kapiti; but Te Ati Awa were left with the uneasy feeling that they, too, could be wiped out, when Tainui chose to retaliate, unless they could find a source of firearms. Only shortly before the *Adventure* had anchored off their coast, another heke — made up of allies from Ngati Mutunga, Ngati Tama and other different iwi and hapu to the north — had passed close to Ngamotu. Its members reported recent fierce attacks from Waikato and Ngati Maniapoto. They, too, knew that more of these would completely overwhelm them, and were withdrawing to Kapiti. The villagers at Ngamotu were abuzz with anxious debate about their own prospects.

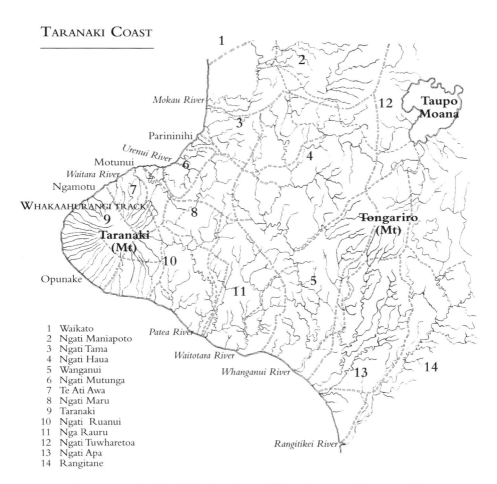

*Approximate tribal boundaries (–··–··–) pre-1830, as they were before most of the people of these tribes migrated south to the Kapiti Coast and Whanganui-a-Tara, away from Tainui's guns. Tribal tracks are shown (---), including Whakaahurangi track, on which Tama-te-uaua began.* ALEXANDER TURNBULL LIBRARY

In the Englishmen's first few weeks at Ngamotu, however, no one sat around waiting for attack. They were far too busy preparing for winter. It being March, they had already dug and stored in large underground pits most of their kumara, taro and hue. Now it was time for fishing from canoes, diving off the rocks for koura and other shellfish, and shooting and trapping edible birds in the great forest behind them.

Some of their pickings they ate, some they preserved. The men repaired and made fishing nets and gardening tools, and fashioned clubs and spears for fighting; the women wove mats for clothing or plaited kete. There was no spare time at all.

The seamen must have watched and adapted and learned; and the tribe adapted too, cutting and dressing harakeke (flax) in large quantities for Love and Barrett's principals. By tradition, every family member learned to prepare muka. It was the basis of their clothing and baskets, but it was time-consuming and labour-intensive.[12] The villagers were initially motivated by the prospect of the guns they would acquire; but, since the going rate for one good gun was 406 kilograms of dressed flax, and for a 22.7 kilogram case of gunpowder it was 1.02 tonnes of the product, spirits and quality soon began to flag, since a 'good hand' could produce only 5 kilograms a day.

One night, soon after the *Tohora*'s return from that first trip to Sydney,[13] the ship's anchor dragged during an unexpected sudden gale, and she was driven onto the village beach at Otaikokako, near her anchorage. She was not unduly damaged. Unfazed, teams of Ati Awa emptied the rest of the Australian cargo from her holds and, after minor repairs, hundreds of them made a slipway of small branches covered in seaweed and hauled her back into the water. However, what started as a minor mishap ended in calamity. As they began loading her with fresh goods, a cask of pork fell out of the slings and crashed through the hull. The almost-new ship was scuttled and the Englishmen were stranded and isolated.

This happened towards the end of June. Yet, happily, by 19 August the marooned mariners were crossing the Tasman to New South Wales in the brig *Elizabeth*. The trading vessels plying across the Tasman Sea and up and down the New Zealand coast ensured constant if often slow communication between New Zealand and Australia. Either the captain of the *Elizabeth* heard of the marooned seamen's plight when he reached New Zealand, or Hyndes and Street received a message about the accident via another vessel, and requested help from the owners of the *Elizabeth* before she sailed from Sydney for New Zealand on 19 July. Whatever the order of events, the crew of the *Tohora* were back in Sydney to report to their employers on 9 September 1828 — but telling everyone else there that they had been rescued from an abortive sealing voyage. They were making sure that no one else realized the potential of New Zealand flax!

It would seem that only Love, as senior member of Hyndes and Street crew, needed to return to Australia, to discuss alternative shipping with his principals, but all the English seamen were on that boat to Sydney. They may have envisaged returning to 'civilized' society for good and leaving their Ati Awa partners as just temporary wives in the port of Ngamotu. But the attractiveness of living among 20,000 people, streets, docks, houses and shops, after being ensconced for six months among friendly Ati Awa tribespeople, may not have measured up as they expected. Wakaiwa and Mereruru were pregnant when Barrett and Love departed and probably the other men left behind pregnant wives, too. Within a month the whole crew of the *Adventure* was back at Ngamotu.

Apparently, there was a split in whatever partnership had existed between Hyndes and Street, and arrangements had been made for Ngamotu to be serviced for the next few years by ships owned or leased only by Hyndes. His own vessel, the *Admiral Gifford,* made two trips in the following five months, to collect their goods, then the trader *Currency Lass* carried their flax and other products on consignment. In time, however, other trading ships called at what became an increasingly well-known trading centre.

Everyone there worked extremely hard and Barrett, who became the business manager of the project, had much to learn about the tribe's farming methods. The areas of fernland indicated where old cultivations had impoverished the land. They were never used again. Each season, about two months before sowing should begin, the villagers left their homes in small parties, walked inland a short distance and selected further areas of forest for that year's fresh cultivation patches. Each group felled their patch's growth (except for the karaka trees, which were prized for their fruit), dried, burned and sketchily cleared it, then attacked the exposed earth, loosening it with ko (pointed sticks) before inserting seed. Previously, the tribe had grown crops only for themselves, or guests, or for gift exchange.

Now, Barrett introduced novel seeds from Australia, and persuaded Ati Awa to prepare, sow, tend and harvest considerable areas of melons, maize, cucumbers, pumpkins and wheat in their cultivations and around his home, and to raise extra numbers of pigs for the export market.

There were problems, of course. First, there were the pigs. They were a vital part of the trading system, once they were bacon; but on the hoof they were a constant nuisance as they discovered the attractive new shoots. Many more pig-proof fences had to be constructed. Secondly, by a tradition which Barrett could not alter, every capable person in the village took part in planting and harvesting the crops; therefore flax processing ceased from October through to December that year, and would again every following February and March. The third problem was that that entire period was also the time for repaying old debts, gathering in more slaves, and training the unfledged warrior-striplings of the tribe. The whole summer was thus effectively lost for flax-processing.

This became apparent soon after their return from Australia, when a large band of their fittest gardeners downed their ko and headed off south, flourishing their taiaha, mere and patu. They were going to join other Ati Awa warriors intending to fight alongside Ngati Toa in Te Wai Pounamu. Te Rauparaha, who had established himself as the most powerful chief in the Kapiti area, had designs on this large island across the straits. The young men, overflowing with vigour and ambition, looked forward as eagerly to the annual fighting season as they might do to the rugby season a hundred years later.

It was carefully planned by their more experienced chiefs, who aimed to achieve tribal or family utu for previous deaths and defeats, as well as tribal and personal mana. It seemed unwise to their English partners that, with the possibility of attack from Waikato and Ngati Maniapoto imminent, Ati Awa sent valuable warriors so far away. The reason, they learned, was utu in another form. Ngati Toa were historical allies, and Ati Awa helped them when they could, even if it was slightly inconvenient. So how long would the fighters be away? When the crops were due to be harvested they should return, would have been the reply.

And most of them did, tired and often wounded, but full of tales of battles and exhilarated. Some had been present when Te Peehi Kupe, hereditary ruler of Ngati Toa, had been killed at Kaiapoi by the 'treachery' of Ngai Tahu's principal chief, Tamaiharanui,[14] and spoke with anticipatory awe of Te Rauparaha's promised vengeance. They spoke also of Te Rauparaha's grandiose ambitions to conquer the whole of Te Wai Pounamu. The remainder of the gardener-warriors had stayed with their Ngati Toa fighting friends across the strait. They were cultivating Te Rauparaha's conquests on his behalf, so that in three years' time he would be able to lay legitimate claim to it. By hearing a snippet here, a boasting tale there, Tiki may have gradually absorbed a touch of tribal land lore.

At this time, with most of the warriors back, and preparations for the second planting season under way in the spring of 1829, Barrett began looking for more trade. Te Puni and Te Wharepouri were senior directors in charge of labour, Love was a sailor at heart, the captain of their sadly defunct ship and always the seamen's basic leader on land, but not an entrepreneur at heart. Barrett was, and he travelled constantly and exhaustively, becoming a well-known trader as he built up contacts, seeking deals in flax and guns to funnel through Patarutu.

On foot, and probably accompanied by Ati Awa guides, he followed the native tracks. He prospected north along the coastline to Mokau and beyond. Covering that country was tough going. At places there was no beach and the track went up precipitous cliffs, frequently intersected by ravines and with forest growing right to their edges. Now Barrett found himself in land belonging to other tribes — Ngati Mutunga and Ngati Tama. They were allies of Te Ati Awa and also interested in trade. Going east, he tramped for days in the sombre green, filtered light of the thick forests at the back of Ngamotu,[15] where interconnecting valleys of the Whanganui River and its tributaries gave him access to the Bay of Plenty. He even established a monopoly over flax supplies right across at Ahuriri, in Hawke's Bay. He went southwards to barter with Moki Te Matakatea and the not-always-friendly Taranaki tribe.[16] He may even have gone on from there and dealt with a flax trading post at Whanganui-a-Tara.[17]

That terrain was easier. The forest stood 3–6 kilometres from the coast, the cliffs were lower, the beaches more frequent and the population very numerous. As the husband of Tautara's granddaughter, he would have been given automatic mana, but his friendly personality would have gone far to assure him of an enthusiastic welcome at most of the villages he visited.

Eighteen-twenty-nine was a vintage year, with ever more new experiences. With Wakaiwa — whom Barrett now called Rawinia[18] and sometimes Rangi — and Mereruru pregnant, both he and Jacky may have found themselves collecting nikau fronds and flax, and totara or kahikatea bark, to help build special whare kohanga. These small, temporary structures for the birth process were erected when the mother-to-be was about seven months' pregnant. At 2.5 metres by 2 metres and 2 metres high, they were cramped conditions for husbands to be supportive in; but usually the young woman's mother attended her.[19] In February, Rawinia produced a daughter, Kararaina, or Caroline, and Mereruru a son, Hone Tanerau, or Daniel, about the same time.[20] The births would have occasioned much feasting and rejoicing, along with destruction of the whare kohanga, now tapu after the birth process.[21]

Ngamotu's reputation now was steadily growing as a small civilized centre for trade and local produce. Barrett and Love had also introduced the tribe to exotic peach tree seedlings, brought from Sydney in the *Admiral Gifford*, or *Ameriki Watiti*, as it became known locally.[22] Numbers of trading ships belonging to other Australian merchants began to call in.

In both the 1830 and 1831 fighting seasons Barrett's gardeners disappeared on active service again; and, once more, some of them never returned. They sent messages for their whanau or hapu to join them, and settled into Kapiti, or Whanganui-a-Tara, or on Te Wai Pounamu. Here there was something novel offering — work with English whalers, one of whom was called Guard. Those who came home brought exciting stories of the campaigning, and of the crush of people now living on top of each other in the Kapiti region as a result of the many migrations. Waitohi, Te Rauparaha's influential and highly respected sister, had earlier firmly assigned the various hapu there separate areas in which to live, but this had not prevented friction, especially between Ngati Raukawa and their Ati Awa kin living round Horowhenua Lake.

In 1831, Te Wharepouri, during his accustomed travels, found inviting, unoccupied land at Wairarapa. Ngati Kahungunu, its previous resident iwi, had retreated north and were living mainly towards Mahia, he reported. They had apparently been squeezed out of their homeland by migrants flooding southwards, most of whom were Ati Awa allies. If Te Rauparaha had moved Ngati Toa to Kapiti, to avoid nemesis, Te Ati Awa could move to Wairarapa, if necessary, Te Wharepouri suggested, and again live next door to friends.

Notable also in 1831 was Te Rauparaha's taking expected revenge on Tamaiharanui with the help of a conniving English sea captain; but of this Barrett may already have been uneasily aware. The brig *Elizabeth*,[23] which had been involved in the grisly matter, had been under lease to his old principal, Thomas Street, who had sent it to New Zealand seeking a cargo of flax. Its captain, John Stewart, made a pact with Te Rauparaha and, for 50 tonnes of flax, carried Te Rauparaha and about 100 warriors to Akaroa. Here, apparently, Ngati Toa, possibly helped by the sailors, sacked and burned Tamaiharanui's village. That chief was among other captives brought back to Kapiti, held prisoner on board — this time definitely with the connivance of the crew — and later slain and eaten.[24]

One account of the affair says that Te Rauparaha then reneged on the flax payment to Stewart, producing only 16 or 18 of the 50 tonens promised, while another story indicates that the whalers at Kapiti were so incensed against Stewart that he had to flee before his holds were full. Yet the *Elizabeth*'s hold contained 30 tonnes when it berthed in Sydney on 14 January 1831. It is possible that, either way, Stewart topped up his cargo at the nearest flax centre, Ngamotu, in which case it seems improbable that her sailors did not talk about their experiences with Te Rauparaha. If so, and if Stewart wanted flax, Barrett had a dilemma. Firstly, there was trade involved — that should never be turned down. Secondly, Te Rauparaha's affairs were none of his (Barrett's) business. Thirdly, the ship was travelling on Street's affairs and Barrett may have felt obliged to co-operate. However, if he did, he would be co-operating in a reprehensible deed. Repercussions of 'the *Elizabeth* Affair' jolted immediately as far as Australia, and large waves, more than ripples of the affair, disturbed Downing Street.[25] Apparently, neither Hyndes nor Love nor Barrett ever again had anything more to do with Street, the impression being that the one joint operation with the man was one too many.

Te Rauparaha's initiative in this affair was quite within the bounds of clever tribal tactics. Although he was roundly condemned by Europeans, who called him 'The Old Sarpint' — among other less complimentary sobriquets — he was feared by iwi all over the country, yet admired by them all. He became almost a legend in his own time. Vassal chiefs in Te Wai Pounamu sent him supplies for his barter with Europeans, and captains of both trading and whaling ships flocked to his harbours at Porirua and Pukerua, to be alternately charmed and bullied by him. Indeed, Barrett and Love benefited, when these ships often called on Ngamotu afterwards. They could not have avoided being aware of Te Rauparaha's formidable presence and effectiveness.

During 1831 the Barretts' flax profits were diminishing sharply, but momentary welcome relief came with the birth of their second daughter — Dicky's 'lovely child' Mereana, or Mary Ann. Profits from flax were dropping all round New Zealand, in fact, as the result of an unexpected vicious circle. The early supplies of high-quality flax had brought many guns to many tribes. The firearms had led to more fighting and more casualties, resulting in there being fewer people and less time to dress flax. As a consequence, further supplies of flax were prepared hastily and badly, merchants and the British Navy became disenchanted and demand tailed off; but the tribespeople did not much care. They had their firearms. The trading post's original *raison d'être* dwindled; but before it became defunct, and while Patarutu was still full of flax, word flashed around the region in December 1831. Waikato were coming.

*Chapter 2*

APRIL 1832–JUNE 1833
SIEGE, MIGRATION AND HARDSHIP

Undergoing an immediate metamorphosis, the Englishmen became part of the tribal defence force. They helped as kumara, taro, and the small crop of potatoes grown were hastily lifted from the soil and stored in pits within the palisades and as the villagers deepened ditches they threw up bullet-proof banks of clay and fern around their whare and stockpiled other provisions. Although Otaka pa, standing on a steep rise just up from Otaikokako beach[1] and covering less than half an acre, was puny compared with the Pukerangiora pa at Waitara, Ati Awa looked to it as a defensible bastion.

In hindsight, Ati Awa wondered whether two Waikato canoes that had landed at Waitara several weeks before had carried spies. Local people had helped repair the craft, and sent them off filled with provisions and the special Taranaki dried fish which their occupants had sought. In retrospect, these Waikato could easily have learned on this visit that there were now no tribes occupying the country between Waikato and Te Ati Awa. Ngati Mutunga and Ngati Tama, the buffer tribes, fearful of their neighbours' muskets, had been part of more than one heke moving south away from their tribal lands. As well, as mentioned, many Ati Awa had already emigrated to Kapiti and Te Wai Pounamu, and some of their remaining warriors had recently left temporarily to join Te Rauparaha's latest expedition. Waikato could have suspected the vacuum, and the spies could have confirmed it would be an auspicious time to give their young warriors fighting training and to avenge Motunui.

By 24 December 1831, the seamen and Ati Awa could see the fires of the Tainui taua up the coastline; and by the beginning of January 1832 refugees were turning up at Ngamotu. They were the remnants of the thousands of Ati Awa who had still been living on their tribal lands further north when the Tainui force invaded. Normally, they would have defended their homes; but not this time, when they had seen thousands of guns glinting in the sun. Abandoning their crops, some had fled south, while others had poured into the strongest pa in the area — Pukerangiora.[2] Previously, this fort had been impregnable, placed as it was at the top of steep cliffs and hundreds of feet above the Waitara River; but in this emergency it was under-provisioned and the people were disorganized. Their chances of holding out for long had been nil when the Tainui taua settled down to a siege.

The Englishmen heard tales of horror from the refugees. Ati Awa had tried to break out, after two or three weeks cooped up in Pukerangiora with little food and no hope. They had been shown no mercy. Over 1,200 had been killed or captured, out of the several thousands who had been trapped inside. Te Wherowhero had ordered that prisoners with fine tattoos be carefully decapitated, to provide mokomokai for trade.[3] Others were simply slaughtered and prepared for the umu. The only prisoners to survive were the ones kept as slaves and possible further meals. Among those who had escaped, apart from those now at Otaka, some hundreds aimed to reach Waikanae and Kapiti, others to seek protection among Ngati Ruanui, the next tribe to the south, while a few had melted away into the forest around the base of Taranaki mountain, hoping to eke out an existence close to home.

Clearly, the prospects of the people in Otaka, if they were defeated, would be similar; but the chiefs there had no thought of retreat and continued with last-minute defences. The tetere was blown, its moaning notes requesting reinforcements echoing over the countryside, and the Ngamotu fortifications were optimistically extended to contain Mataipu pa nearby and the village canoes and the refugees still streaming in.

The Waikato messengers who turned up at Ngamotu at the end of January represented emissaries of a split decision. After the post-battle feasting at Pukerangiora, most of the Waikato chiefs had felt that Motunui was sufficiently avenged and wanted to return home with their slaves.

*Te Wherowhero, who threatened to steam and preserve the heads of Barrett and Love and the other Europeans (Seffern p. 10), and who later became King Potatau, the first Maori king.* ALEXANDER TURNBULL LIBRARY

One Ngati Maniapoto chief, however, still felt that 'revenge had not yet been satisfied', and had advocated attacking Otaka pa and the remnants who had escaped from Pukerangiora. His taua colleagues reminded him that the men he was now intent on destroying had helped him in an earlier battle;[4] but, after discussion, they were outvoted and the taua had started out towards Te Puni, Te Wharepouri and their warriors.[5]

Otaka by now was overcrowded. More fugitives flooded in every day, as well as reinforcements from Pukeariki, Pukaka and other pa close by, responding to the tetere, and tribal members from further afield, who had heard of the invasion. Some of the latter, including Ngatata-i-te-rangi, a venerable chief of Tautara's age, and his son, Wiremu Tako Ngatata (commonly known as Wi Tako)[6] had returned to defend their home pa. Others took up defensive positions in pa on Mataora,

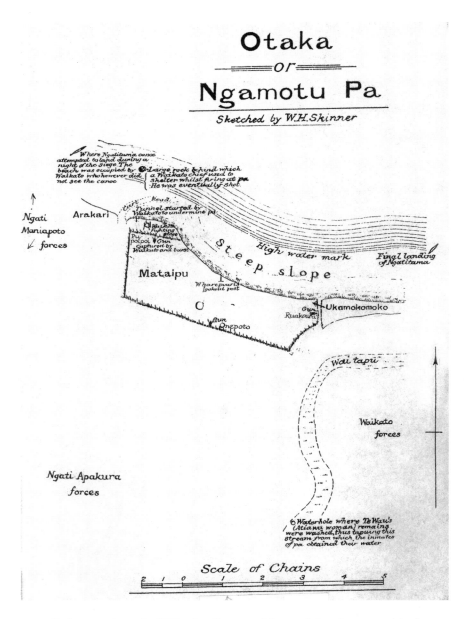

*Otaka pa, as it was in 1833, when Barrett and Love and the other seamen joined Te Ati Awa in fighting for their lives against surrounding Tainui forces.*
ADAPTED FROM SMITH, MAORI HISTORY OF THE TARANAKI COAST

Motuotamatea, Mikotahi and Paritutu. The elderly and other defenceless members of Ngamotu village were placed there for safety, too, including, probably, Rawinia Barrett and Mereruru Love and their babies.

The seamen, meanwhile, trundled their three heavy cannon into strategic positions inside the pa.[7] Typically, Ati Awa had given them Maori names. Ruakoura commanded the country on the east side of the pa, Onepoto faced inland, and Pupoipoi, a small field piece on wheels able to be moved if necessary, occupied the north-west angle of the fortifications.

There are conflicting reports about what followed,[8] but Barrett, Love and the seamen found themselves involved in three stressful, February weeks of war, run on lines supremely different from Western battle conventions. All accounts agree that Te Wherowhero, the Tainui leader, made the first move. He was a tall warrior of magnificent physique, famous for his feats all those years ago at Motunui where, single-handedly, he had fought and disabled dozens of his enemy. At Ngamotu, he appeared on the beach waving his mat — a sign that he wanted korero — whereupon senior chiefs from both sides gathered to meet on the black sand. There they spent time leisurely reviewing past friendships, family connections and battles against mutual opponents, and mulling over peace proposals.

Then the accounts begin to differ. Tainui said that Te Wherowhero and Te Wharepouri decided on an armistice for several days, at which, in relief and jubilation that hostilities had been averted, '. . . the Waikato . . . did a haka in front of the fort, discharged their guns [as a salute], threw them down [for their women or slaves to retrieve] . . . and rushed forward and up the hill to the pa'[9] to hongi with their relatives — only to be mown down by gunfire from within. Their bodies fell everywhere and, in dismayed retaliation, 'the amazed besiegers [who had not yet thrown down their guns] rushed past their slaughtered people . . . and fired shots at the pa with some effect'.[10]

The seamen's story, on the other hand, reports that all the palaver seemed to be an attempt to attack the pa by bluff. They could not afford to believe that this was just a boisterous, peaceful visit. With their babies and homes apparently threatened, they fired guns and cannon in a gut reaction. The supposed armistice disappeared in a blaze of indecisive battle before Tainui withdrew to their camp.

There followed a period of non-engagement, filled with impromptu haka by both sides, the digging of saps, and even a fairly good-humoured wrestling match between two giants. This interim was broken by a surprise early morning Tainui attack when their troops suddenly appeared out of nowhere and dropped down inside the pa. They had felled karaka trees, growing on the edges of Otaka's defensive trenches, propped them up against the palisades, and used them as assault ladders. There was confused fighting with casualties on both sides before the attackers were driven out.

Sporadic attacks continued, alternating with occasions when both sides met and tried to come to an agreement; but the Tainui reports about these say that the Englishmen were the ones who turned down their peace offers and insisted that they must 'fight to the last'.[11] while the Europeans quote continual suspicion of Tainui's supposedly treacherous motives. For whatever reasons, the peace offers were refused and the siege continued.

Descriptions of the following period range from minimal and matter-of-fact to lengthy and lurid, but it produced anxiety, high-adrenalin levels, and constant exhaustion among the white men. It was the first time they had literally fought for their lives, and they were often bewildered by the many aspects of the conflict. One of Te Wharepouri's brothers, Tohukakahi,[12] jumped the walls to make a suicidal and apparently pointless solo foray right up to the enemy and was killed. Another Ati Awa chief accepted a challenge to engage in an individual fight to the death outside the pa. He won, but it made no difference to the general situation.

At night, the Ngamotu troops slept tranquilly, wrapped securely in their blankets and in their knowledge that, as custom prescribed, the enemy would not attack during the hours of darkness. Barrett and Love, however, became worn out after deciding that someone had to remain alert twenty-four hours a day. They were astonished to see friends and

SIEGE, MIGRATION AND HARDSHIP

*An imagined depiction of Dicky Barrett at the defence of Ngamotu pa, February 1832, by A. H. Messenger, an illustrator for James Cowan's* The New Zealand Wars.
JANET WILSON/TARANAKI MUSEUM.

relations from both sides meeting and chatting between attacks, and wandering freely and unmolested between the two camps.[13]

They were amazed and dismayed, after repelling several assaults, to see several important Tainui chiefs entering the pa to share a meal and be 'politely' shown the defences.[14] This resulted in those 'guests' exhorting Ati Awa to make peace and surrender. The defenders in the pa were hopelessly outnumbered, they said; but the Englishmen declined, outraged at having their weaknesses displayed. Again, there were different interpretations of this affair. One saw it as a genuine move from Waikato — that they were going to free all prisoners and go home — and suggested that its rejection arose from the Europeans' imperfect knowledge of Maori language and habits. Another said that Ati Awa would probably have accepted the proposal to surrender 'but for the English', who suspected treachery.[15]

Whichever interpretation is correct, the Englishmen had fresh surprises ahead, when they discovered that there was bartering going on between the antagonists. Waikato and their allies' guns outnumbered those of Te Ati Awa to an extreme degree, but the travelling taua was short of supplies and hungry. As Ati Awa had plenty of tobacco, other articles of trade, and quantities of food, a brisk business took place as guns were swapped for provisions.[16]

Meanwhile, the besiegers dug more saps to undermine the palisades and to avoid the cannon shot, and built towers from where snipers could shoot over the defences, and from where flaming brands could be thrown onto the thatch-roofed whare in the pa. The defenders, in turn, dug counter saps, traded musket shot for musket shot with the marksmen up high, and doused the flaming torches before the flames took hold on their dwellings.

In these ways, Ati Awa were effective fighters. In other ways, they

*One of Barrett's cannons, bought from a passing ship and used during the siege of Otaka pa. Reputed to be an antique, even at this stage, it is now at Taranaki Museum.*
TARANAKI MUSEUM

made Barrett and Love despair. During the day Te Wharepouri often stood on a raised platform, with a view of the foe, in a position to command proceedings. He was strong and sinewy, vibrant, dominating and headstrong, and his men looked to his leadership; but many of them shared his characteristics, and individual chiefs persisted in carrying out their own sorties, like Tohukakahi, if and as they wished. The pa seemed a bubbling confusion of noise and movement to the Europeans, with much shouting, a continual discharge of guns, boasting about past exploits and bragging about brave ones to come, and quarrelling over tactics. On top of that, neighbours still squabbled with one another, and family members continued with domestic disagreements in the middle of their common danger. Sheridan reminisced later:

> We, the white people, have frequently been in more dread of the natives in the par than those outside, expecting civil wars amongst them. . . . If any of their relations happened to be shot, there was sure to be a row, allusions to different frivolous faults; so that we were constantly busy trying to keep peace inside, and look out sharply for those outside.[17]

On one occasion, two sisters quarrelled when one informal peace korero had ended in stalemate, and the one who had sided with Waikato raced out of the pa, to be with those she had supported. They gave her short shrift, however, and killed her on the spot; then, food being scarce, cut her into pieces in full view of the defenders, washed her carcase in Hongihongi stream (thus putting a tapu on the water which supplied the pa), and cooked and ate her. The defenders were able to dig alternative wells and did not suffer, but the Englishmen saw it as pure Tainui savagery.[18] An early settler by the name of Jervis, on the contrary, explained the killing from the Tainui point of view, as inevitable retaliation for a previous breach of a truce by Ati Awa.[19]

During the second week of conflict, the trading schooner *Currency Lass* appeared off the coast. The sight of the 90-tonne vessel could have inspired hope in the traders, but the visit ended in an impasse. Captain Buckle[20] kept his ship outside the range of musket shot and provided neutral decks for the meeting that Te Wherowhero requested. The chief went by canoe. Love, as Otaka delegate, swam to the boat.

Before the emergency, Love and Barrett had had Patarutu full of muka, ready for the next trading vessel that called. Now Love heard Te Wherowhero baldly offering this flax to Buckle, in exchange for the goods in the *Currency Lass*. His warriors had found it in Patarutu and it was spoils of war, Te Wherowhero said. His warriors were starving, and if the captain would pass over his ship's goods, he, Te Wherowhero, would in return end the siege and spare the Englishmen's lives — although, of course, he would take them back to Waikato as slaves.

Love knew enough about Maori war tactics to realize that there was a fifty-fifty chance that Te Wherowhero would be true to his word. The odds were not good enough. He objected vehemently, and Buckle refused Te Wherowhero's offer; whereupon Te Wherowhero reverted from suppliant to aggressor, intimating '... an immediate occupation of the pa' and that the Pakeha heads '... would be steamed and preserved for sale....' Love retorted that 'they were perfectly satisfied with the position their heads occupied....'[21] and swam back to shore through a hail of musket shot, while the *Currency Lass* sailed away.

A night or two later, thirty or forty Ngati Tama allies of Ati Awa, under the leadership of Te Kaeaea (later to be known as Taringakuri), landed from a canoe on Otaikokako beach. They eluded Waikato surveillance to slip into the pa as welcome reinforcements, but the boost they gave to Ati Awa morale was somewhat shaken soon after when the little swivel cannon exploded.[22]

By now, the last week of February was approaching. The taua's food had run out and Tainui had just heard that other enemies had invaded their homes at Mokau in their absence. They had had enough of this protracted struggle in alien territory. They wanted to end it and see to their own home defences. Te Wherowhero took the accident with the gun as a good omen. He announced to those in the pa, in the customary fashion, that he would mount a final assault the following day and annihilate them. As promised, Tainui attacked the next morning, breached the palisades and poured in.

Again, there are two versions of this assault. The seaman Sheridan reported that the attacks on three different sides were unco-ordinated; and that, although the ratio of at least five Tainui to one Ati Awa seemed insurmountable, the belching cannon, coupled with the inspiration and leadership of Barrett and Love, repelled and almost literally shattered the invaders. This is supported by one historian's report, which added that the seamen Bundy and Wright were highly admired, Tiki was a 'tough little warrior', and Hakirau became known as 'the better general'.[23] By contrast, Jervis indicated in his account that Ati Awa tolerated and somewhat ignored Barrett and Love as fighters, rather than seeing them as inspired leaders. He intimates that, at this point, the Waikato allies ceased trying to make peace, or even to subdue Otaka.[24] After fighting in the pa for a while, they withdrew. Their priorities were now at Mokau. They put their dead on the roofs of the temporary whare they had built during the siege, and set fire to them, so that they did not become calories for their enemies. Then the warriors started the homeward march, taking with them the slaves they had captured at Pukerangiora, satisfied with their gigantic victory there, but irate over their losses at Otaka, which they gave as 'four chiefs of rank ... and sixty of lesser rank'.[25]

Sheridan's account said the Tainui tribespeople fled, and that Ati Awa chased after them, inflicting two hundred or more casualties. This is certainly an exaggeration, but obviously there were heavy casualties during this engagement. Jervis was one of the settlers who found human bones as they dug foundations for a store in New Plymouth in 1841; and, a century later, parts of skeletons could still be found in the sandhills to the west of Hongihongi stream. Waikato had achieved utu for Motunui ten times over; but at a cost which meant that more revenge was still owing.

One history recounts how, to the Englishmen's disbelief, given the extreme shortage of arms among Te Ati Awa, each fallen Ati Awa chief was interred with due reverence and eight muskets, and ammunition to go with them.[26] Then, according to Barrett's journal,[27] the victorious defenders fed their injured and defeated enemies, and gave a pious, Christian burial to their dead. According to other accounts, however, dismembered Tainui carcases decorated Ngamotu doorways, days of celebration and feasting followed, dogs ate their fill, February flies came in clouds, and Barrett, Love and the seamen could only look on in revulsion, thankful that it had not been their fate.[28]

In the triumph and relief of the moment, the Englishmen had not considered the implications. The people at Ngamotu knew only too well. Retaliation by Waikato would be only a matter of time — utu could not be avoided and David may not again defeat Goliath. They would have to migrate south, like all those other thousands, to Kapiti Coast, or Whanganui-a-Tara, or Totaranui. They would link in with whalers and traders and get guns. Then they would return.

But, while they were still within easy striking distance and before they left, several chiefs from Te Ati Awa and allied tribes decided to settle old scores against Ngati Maniapoto and retaliate for the recent incursion. While the heke was assembling, a strong ope darted up to Mokau on a swift raid. Meanwhile, the partnership at Ngamotu used this hiatus to dismantle the trading post, and to tell several visiting trading schooners about its closure.

Most of the seamen now melted away.[29] Sheridan, for instance, found a passage to New South Wales in one of the schooners. The Barretts, Loves and Keenans decided to go with the tribe. They had gone to Sydney, after losing the *Adventure*, but had returned to Ngamotu. Now, four years later, after losing their entire business, their Ati Awa ties had become even stronger. In New South Wales, they would just be British seamen looking for jobs, as they had been in 1827–28, but now with wives and children as well. Elsewhere in New Zealand, they would have a struggle to find some niche in which to keep themselves and their families alive; but while with Ati Awa, Barrett and Love were honoured both for their high-born wives and the trade and muskets they had engendered, and there would be food and shelter and support for them, and for Keenan and his family as well.

The ope was not away for long. Te Ati Awa suffered few casualties, but achieved no great result either, except to create more animosity and much topic for conversation, as the returning fighters joined everyone assembled near the Waitara River to begin their long walk. A skeleton crew of nearly 300 people, including Rawinia's parents and Poharama, was to remain at Ngamotu to maintain Ati Awa ownership. Everyone else joined the heke, which comprised a band of almost 1,400 warriors and a similar number of old people, women and children. It was probably the largest heke in Maori history that included Englishmen.

Everyone had packed pikau with basic necessities, and they set off in late March like a long, long caterpillar. Principal chiefs Tautara, Rawinia's grandfather, Te Puni, and Te Wharepouri led the column. Behind them, each hapu formed a separate segment, with warriors at the front and back of each, guarding the vulnerable ones in between. A few men were detailed to see that the distance between the groups was maintained and that no straggling occurred. Over a dozen iwi and hapu groups, including some Ngati Mutunga and Ngati Tama, Ngati Tawhirikura and Ngati Te Whiti, made up the units of this long, travelling body.[30]

The first section of their trip on the Whakaahurangi track was short,[31] with two very cold, frosty autumn nights in the dense forest; then they were out into the open, covering probably 20 kilometres a day. They passed safely through inimical Taranaki land — perhaps Dicky's earlier business contacts helped — and on into friendly Ngati Ruanui country. By then, it was early June, and the leaders called a month's halt for discussion and planning. Te Rauparaha had had two or three years in which to plan his migration; this heke, which became known as Tama-te-uaua, was planned in only two or three weeks, and the logistics of feeding, sheltering and defending the immense, slow-moving band, were highly complicated.

Initially, when the group set off again, they foraged and fed well: 'The kakas were very fat . . .'[32] but mid-winter found them living off the land with increasing difficulty. The terrain was more level, but less heavily wooded and there were numerous rivers to cross. The first heke members were killed here as they scouted for ao-kai in enemy Nga Rauru country. Cold, rain and mud further hampered the group, as they plodded on through the hostile environment as inconspicuously as possible.

Weeks later, when they reached the barrier of the Whanganui River, they saw some canoes lying unattended on the far side. Several Ati Awa warriors swam across in glee and paddled them back to use as ferries. To the Englishmen it looked like trouble, and it was. The canoes belonged to Ngati Tuwharetoa, a tribe led by one of the most important and respected chiefs in the whole country, Mananui, Te Heuheu Tukino II. He was a giant of a man, renowned for his wisdom, mana, and prowess in battle, whom few would choose as an enemy. He and his brothers, Iwikau and Te Popo, were acting as escorts to a branch of Ngati Raukawa, enemies of Ati Awa, who were also migrating away from Waikato, to be with their Ngati Toa relatives on the Kapiti coast.

Incensed when he returned and found the canoes missing, Te Heuheu joined up with local Wanganui warriors and attacked the heke. They badly mauled members of Te Puni's Ngati Tawhirikura early in the action, who fell back in disarray. The other heke warriors closed the gap, and first one side then the other seemed to have the upper hand in the confusing hand-to-hand fighting which ensued. Te Popo killed a prominent Ngati Tama chief and was in his turn slain. Barrett, Love and Keenan, stripped to the skin, fought for their lives alongside their associates. Finally Ngati Tawhirikura, having regrouped, attacked from the rear, and turned the struggle the way of the Ati Awa allies. Sensing this battle lost, the enemy jumped into canoes, or into the river, and fled. Ati Awa got into the remaining canoes and gave chase. Amazingly, Ati Awa and their allies on this day were part of another surprising, if possibly short-lived, victory and Wanganui and the great Te Heuheu were routed. This initial engagement took place east of Pukenamu[33] and, in all, the heke lost only thirty-eight men, eight of these being main chiefs.

As they wound down from these adrenalin-filled hours, the Englishmen were treated to a haunting, typically Maori exchange. During the still of that evening, Te Heuheu's voice came floating across the darkness from the other camp: 'Whaakina mai taku tangata, kowai?' ('Who was your most important fallen Ati Awa chief?')

A principal chief of Ngati Mutunga replied. He looked powerful and brutal, with unusual long hair growing on his neck and shoulders but, in a manner belying his appearance, he keened back a haunting lament in answer, his voice echoing off softly with the same question into the silence. Te Heuheu's lamentation answered, bewailing his brother Te Popo. The exchange was dignified and emotional and the Europeans must have been poignantly touched; though not so moved by the careful preserving of mokomokai going on at the same time in their own camp — Te Popo's among them — part of the victory proceedings among Te Ati Awa.[34]

After the mini-victory the day before, the heke personnel hurriedly prepared defences nearby at Te Karamuramu, against the inevitable Tuwharetoa retaliation.[35] Now more knowledgeable after their experiences at Otaka, Barrett, Love and Keenan helped their fellow-travellers. The urgency allowed no sleep. By the light of heaped fires and hand-held flares, all worked feverishly through the night. They chopped small trees and erected palisades, cut flax leaves and wove them through these uprights to form screens almost impervious to bullets fired from old-fashioned muskets, dug a trench and threw up a parepare.

Apparently, the heke warriors were not at all fazed by the prospect of further fighting, but with so many non-combatants among its numbers, there was general consensus that the whole group was in a hazardous situation. They therefore in a hurry wove raupo sails for one of the captured canoes, and seven messengers sailed off to seek help from their kin at Kapiti, now only about 112 kilometres to the south.[36]

Meanwhile, Ngati Tuwharetoa and Wanganui had taken up a commanding position on nearby Pukenamu hill. Once again, the white men found themselves under siege, with both sides bristling at each other, letting off steam in stirring haka and indulging in minor skirmishes.

The defenders repulsed the odd direct attack, but were basically vulnerable, being both dominated and outnumbered. However, before they had to face any major assault, reinforcements arrived; first, 2,000 sympathetic Ngati Ruanui warriors from Patea, who had heard of the heke's plight, followed by 1,600 from Kapiti, coming in direct response to the messengers' pleas. Heke members fed these extras on kao, aruhe and korito, augmented by the bodies of the fallen enemy, which they had been drying and preserving for just such a contingency.[37]

Several of the young firebrand chiefs wanted to attack the enemy on Pukenamu hill with the much-enlarged force; but, by now, they had been at Wanganui a month, it was August, and, as Tautara pointed out, if they did not press on, they would not be in Kapiti in time to plant their next season's vital crops. Both parties must have reached the same conclusion about planting, perhaps by mutual agreement, because hostilities ceased. Ngati Tuwharetoa and Wanganui melted away and a smaller heke moved on across the landscape again, minus women, children and the old people, who were placed in the looted canoes and guided down the coast by Ngati Toa.[38]

There are no accounts of how many 'civilians' died during the migration. Te Puni's youngest child, Roka,[39] survived, and the Barretts, Loves and Keenans managed to keep their children alive, even towards the end of the march, when they existed mainly on pipi and aruhe. However, once they reached Whangaingahau pa, near the coast, earlier emigrant Ati Awa friends and relations welcomed them with a great feast, and food worries were past.

It was January or February 1833 when they reached their ultimate goal on the coast at Kapiti and the heke leaders paid a courtesy call on Te Rauparaha. This was almost certainly the first time that Barrett met this forceful individual, who would have an indirect influence on his future.[40]

In 1833, Te Rauparaha was probably into his sixties, a short man, with a retreating, wrinkled brow, aquiline nose and wide mouth with overhanging top lip. He had eight wives, about fourteen children (the oldest being already in her forties), six toes on one foot, a commanding presence and a formidable reputation as an innovative and relentless fighter. Although inspiring fear through his wiles and unpredictability, Te Rauparaha also commanded respect, because of his skills in war and regard for tribal ethics and etiquette. This he demonstrated as he welcomed Te Puni, Te Wharepouri and other leading chiefs in the heke as allies and as rangatira, and assigned the various hapu specific areas in which to settle. At last, with eight months of trekking behind them, the Englishmen could start constructing new homes for their families.

However, having left behind the security of their old-established tribal boundaries, everyone in the heke moved with uncertainty into the flux that was Kapiti. It was richly fertile land, '. . . birds were prolific, eels abounded and fish were plentiful . . .' but people were over-plentiful.[41] Te Rauparaha had created problems for himself when he stimulated so much migration. Hundreds of whanau from dozens of hapu and iwi had arrived over the past few years, and were physically and psychologically affecting one another, as they shared out between them the now crowded area which, only fifteen years before, had been sparsely inhabited. These other iwi and hapu who had followed Te Rauparaha were all allied to, or related to, Ngati Toa, but not necessarily to one another. Within the new population, gathered under Te Rauparaha's leadership, were many with old scores still unsettled among themselves; while prickling against Te Rauparaha himself were previous animosities he had incurred from the tangata whenua[42] during the years 1819–20, when he had been there with the northern taua. The resettlement of the tribes, moving into the southern part of Te Ika a Maui, was bringing turmoil.

When the Barretts, the Loves and their Ati Awa heke companions settled at Te Uruhi — the area designated for them by Te Rauparaha — they found that their exploits had given them a new identity. They had become known collectively as Ngamotu hapu. This the Englishmen came to understand, but they were absolutely baffled about who their neighbours were.

The new immigrant tribespeople referred casually and inaccurately to any of the previous tangata whenua living nearby as Ngati Kuhungunu, which is what Barrett came to know them as too; but that was incorrect.

They were Muaupoko, Rangitane, Ngati Ira and others, as well as Ngati Kuhungunu, and some of them had been living round the shores of Raukawa Moana (Cook Strait) for many generations. However, these tangata whenua did no better, referring to all the new 'intruders' as Te Ati Awa. Those on both sides who were not properly recognized were indignant. Barrett must have been confused.

As well, he and Love faced an unsettling situation. Although they may have acquired admirable reputations as fighters, their position within the tribe could not have fulfilled any in-built need for challenge and achievement. They were no longer the esteemed Pakeha traders, in partnership with the principal chiefs of the area. Te Rauparaha was now the influential trading leader. Their four years' experience in Taranaki faded beside Te Rauparaha's entrepreneurial stature at Kapiti, and they would be very subordinate to him. His skilled generalship was matched by his extensive knowledge of horticulture. He moved around constantly, supervising chiefs and hapu working on expansive cultivations at Kapiti, Pukerua, and Porirua in Te Ika a Maui, and at Wairau and Kauraripe in Te Wai Pounamu. Together with other chiefs, Te Rauparaha actually operated in a Mafia godfather style, encouraging trading, and exuding amiability and 'protection' to the whaling communities under his sway, at the same time subtly generating fear among them as to what would happen if they did not co-operate. In this way, he blackmailed them into providing guns and ammunition, European goods and implements that he required, and alcohol, for which he quickly acquired a keen taste.[43]

Trade was now burgeoning, and increasing numbers of European ships were seeking the safe anchorages he offered, as well as the tattooed heads and slave women being sold for the ships. Could the Englishmen — should they — link into his schemes? They would certainly have more chance of contact with Sydney living near Porirua than they had ever had at Ngamotu.

While considering their options, Barrett and Love found themselves to be in the right place at the right time. Several separate incidents occurred that opened up the niche they needed. The first occurred just after Te

Rauparaha had assigned land to the heke chiefs. In the manner of the new grand seigneur of Te Wai Pounamu, he then set off from Kapiti on a hunting trip to trap putangitangi at Kapara Te Hau. Instead, he and his party were themselves trapped by a taua-iti, led by Ngai Tahu chiefs. This had been sent as a forerunner of a larger force the following year, with which these chiefs hoped to achieve Ngati Toa's defeat and Te Rauparaha's demise. Escaping the ambush with ignominy and leaving behind many dead, Te Rauparaha fled to Port Underwood and sent hurried messages to Kapiti for reinforcements.

The second incident arose out of the first. Frustrated at missing Te Rauparaha, the Ngai Tahu warriors cast around aggressively for any other Ngati Toa in the Cloudy Bay area. They found and launched an attack on a group working at John Guard's whaling station at Kakapo Bay, where some Ati Awa men were also employed. These tribespeople simply scattered into the bush at the appearance of trouble, but the alarmed white whalers beat a wholesale retreat, hurriedly pushing their boats into the water and rowing in panic for Kapiti. Their arrival, soon after Te Rauparaha's messengers, was the third and crucial event for Barrett and Love.

A few days later John Guard, who had been absent in Sydney when the attack occurred, returned to find his station looted and burned out, whereupon he sailed straight on to Kapiti. Known to be tough on his employees, he was angry with his men for running away, and rounded them up into his ship, intending to sail straight back to Kakapo Bay, to rebuild at once. At the same time, furious with Ngai Tahu and fed up with yet another attack on his 'property', he decided to quit for good his other whaling station at Te Awaiti. Barrett and Love would have heard all this from the buzzing gossip of the tribespeople in the area.

Forty-year-old Guard had been transported from England for a minor offence while still virtually a boy. He had matured into a black-bearded and powerful man, with a reputation for being difficult, ruthless, quick to anger and not easy to get on with; yet, within this short period, it seems that Barrett and Love persuaded him to let them take over Te

Awaiti. It may not even have been a sudden decision for them. They may have talked about such a move during the heke, as one of their options for the future, since they would have seen whales frequently off the Ngamotu coast.

Guard was an agent for the Sydney merchant R. Campbell jnr, owner also of the brig *Elizabeth*, which had rescued the *Adventure*'s crew when their boat was sunk in 1828. Guard's name probably cropped up in conversation aboard the *Elizabeth*, and Barrett and Love may have kept track of him. If so, they would have heard as early as 1831 that he was using Te Awaiti only during the whaling off-season, as a 'summer residence' and trading post. He preferred Kakapo Bay, since it was much easier to kill whales later in the season from there.

Te Awaiti offered an attractive, less stressful home for the Barretts and Loves, with prospects of a new livelihood — whaling — which might be lucrative for them, if properly managed. Ati Awa friends and relatives were already settled in the region, with whom they could form their own community, instead of being submerged among thousands of strangers at Te Uruhi. Barrett and Love had no long-standing quarrels with Ngai Tahu to complicate relationships, and they knew that Pakeha were sought-after residents among the tribes.

Te Awaiti may have been attractive for another reason to the two men, even if they were under a misapprehension. After Te Rauparaha and his nephew, Te Rangihaeata, had 'sold' Kakapo Bay land in Kauraripe to Guard in about 1831–32,[44] Guard, full of importance and enthusiasm as a landowner, had begun promoting Cloudy Bay, on his trips to Australia, as a superb place for European settlement. Indeed, Hobart's *Colonial Times* reported that 'half the people of Hobart Town are crazy to leave for the new Colony now establishing'.[45] Dicky and Jacky may have moved across Raukawa Moana with the dream that a white population would settle close to them. Whatever the reasons behind their move, in late 1833 or early 1834, Barrett and Love separated themselves from the community at Uruhi and, accompanied by a number of Ati Awa, moved across Cook Strait to Te Awaiti.[46]

*Chapter 3*

JUNE 1833–27 AUGUST 1839
WHALING AT TE AWAITI

———————————

At Te Awaiti — the place Guard had called Fairhaven, or Wickett — Barrett and Love found an environment very different from Ngamotu. They were on Arapaoa, the largest island in Totaranui,[1] looking out on Kura-te-Au, a long, sheltered channel in sharp contrast to the mainland's boisterous western coastline. Opposite them, two kilometres across the water, high ridges leaned out against the open sea, barriers between the channel and the maelstrom where Te Moana-a-Raukawa (Raukawa Moana) met the un-Pacific Ocean. Te Awaiti was one of several side-by-side bays facing east. It boasted clear water, except for small families of rocks surrounding three steep-sided islets standing like pawns off the southern end of their beach.

A semi-circle of gently rising flat land was available for their home sites, beginning at the foreshore and bounded by spreading folds of steep hills full of bird call and covered with any amount of timber for their houses or boat-building or the fireplaces. Seabirds soared and fed at their frontage, where the water, strongly tidal and sometimes, in high winds, surprisingly rough, proved to be thirty or forty fathoms deep; but the many bays on both sides of the channel provided safe emergency anchorages. Marring all this pristine beauty was the nauseating atmosphere of Guard's whaling business. Their beach of thick sand and shell, several metres wide and almost 100 metres long, was littered with charred bones and clogged with old blubber. Putrid smells pervaded everything.

As far as whaling went, the newcomers' prospects could hardly have been better. They were handy to the whales' usual route from the north through Raukawa Moana, so their boats would have first pickings — ahead of Guard and the other stations further south at Kakapo Bay in

## TARANAKI KAPITI COAST, WHANGANUI-A-TARA & TOTARANUI

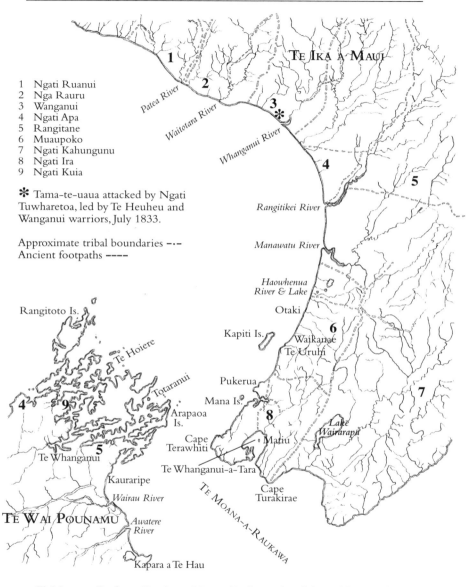

1 Ngati Ruanui
2 Nga Rauru
3 Wanganui
4 Ngati Apa
5 Rangitane
6 Muaupoko
7 Ngati Kahungunu
8 Ngati Ira
9 Ngati Kuia

✻ Tama-te-uaua attacked by Ngati Tuwharetoa, led by Te Heuheu and Wanganui warriors, July 1833.

Approximate tribal boundaries –·–
Ancient footpaths ––––

*Te Moana-a-Raukawa/Raukawa Moana (Cook Strait) and the middle area of New Zealand when Barrett and Love began whaling at Te Awaiti on Arapaoa Island in Totaranui.*
ADAPTED FROM S.P. SMITH, 1910.

Kauraripe — and whales were plentiful.² Te Rauparaha seemed happy for the whalers to work in 'his' waters, and 'allowed' Barrett and Love and Guard to occupy 'his' land. Lastly, two hours away over the hills were assorted groups of Ati Awa friends and relations, who would be ready-made employees. Some of them had moved there recently with Tiki and Hakirau; others had been there for several years, since campaigning with Te Rauparaha against Ngai Tahu.

Barrett, Love and Keenan found themselves joining a small, disparate group of European whalers working on Arapaoa. They were headed by Joseph Toms and Jimmy Jackson, who had both worked for Guard and had opted to try their luck separately as station owners when Guard moved his operations to Kakapo Bay. Toms³ — whose wife, Te Ua, was the daughter of Te Rauparaha's half-brother, Nohorua, principal chief at Guard's station — had been Guard's second-in-command. He had set himself up on the southern half of Te Awaiti, past the second of the little streams for which the bay was named and from which the community drew its water. He was about eight years older than Barrett and a walking example of the dangers of whaling, having been crippled on the job and earned the nickname 'Georgie Bolts'. He was reputed now to flee at even the sight of a whale.

Jackson had his whaling station in the bay next to Te Awaiti — a narrow, spade-shaped indentation across a high tongue of land to the south. Onapapata, his 'snug little cove' was called. He was a cheerful mountain of a man, a non-stop talker, and a draught-playing enthusiast *par excellence*, and had first come to New Zealand as skipper in a ship in 1809, and to the Totaranui area with Guard in 1827. His little cottage was hung with brightly coloured, tinsel-framed pictures of Napoleon Bonaparte's battles, on which he was an authority; but he also had an opinion on anything and everything, backed up by quotations from his two Bibles or his Guthrie's *Geography*. Guard was a part-time member of this European fraternity over the next year or two, when he returned to his old Te Awaiti house for the summer off-seasons.

As far as relationships with other tribes went, the atmosphere could

*Te Awaiti village, known as 'Tarwhite' by visiting sailors, with its whalebone fences. Sketched by John Wallis Barnicoat, 1843.* HOCKEN LIBRARY

have been a lot better. Neither Toms's connection with Te Rauparaha nor Guard's patronage by that chief had prevented the whalers at Te Awaiti from being hassled by passing taua. Barrett and Love could not expect to fare any better. Nor did they.

In early March 1834, a few months after they had relocated, they were attacked for the first time. The long-planned tauanui[4] led by the same three Ngai Tahu chiefs who had failed to kill Te Rauparaha the year before, came north from Otakou intent on completing that mission and revenging the 1831 slaughter of Tamaiharanui. Te Rauparaha, however, had left Te Wai Pounamu for a while: his empire-building had been interrupted when he became embroiled in fighting at Haowhenua,[5] which erupted among the iwi after Barrett and Love left Kapiti.

In frustration, the 400 Ngai Tahu again hunted Ngati Toa, sacking every whaling station in the vicinity and putting to death any Ngati Toa workers they captured. Barrett and Love were not the object of the attack, but it was alarming, to say the least; and then they had to go on working while some of their men's erstwhile attackers lingered around for two months, hoping for Te Rauparaha's return. James Heberley, who, with his wife and two young children had moved from Guard's employ at about this stage to work at Te Awaiti, wrote phlegmatically later, '. . . we lost about forty men which they cut up and put in Baskets to eat . . .'[6]

Despite this set-back, the whale boom and hopes of again becoming part of a considerable European community made prospects for the partners' future decidedly promising. At the start, however, despite the tribespeople providing them with food in abundance, their whaling facilities were fairly minimal: one boat, and only enough gear to enable them to process and send baleen to New South Wales, not oil. Guard, who had taken most of his gear with him, had never built sheerlegs[7] for hauling up blubber from the carcases. His teams had simply pulled up the mountainous bodies at high tide, hammered in supports to stop them slipping back, and put ropes around them to stop them floating away.

Barrett and Love began from scratch, probably watching Toms and Jackson and following Guard's methods as best they could. Barrett may have had some experience in whaling before he came to New Zealand, and he and Love may have caught the odd whale at Ngamotu, but they had minimal training for this new enterprise in unknown southern waters.

For a start, they needed money to invest in proper equipment for processing the whales: boat spades and iron hooks to cut into the blubber and sheerlegs to lift the immensely heavy sheets of it from the sides of each catch; several types of knives to slice these into blocks, then into smaller and smaller manageable pieces; and large try-pots in which to boil this 'mince' over red-hot, wood-fed fires until all the oil had been extracted. They would need other containers as well: tubs to hold the 'mince' while it waited its turn in the try-pots; large, square tanks for storing the oil (and ladles with which to transfer it), and large casks in which to export it.

Above all, they needed a special breed of men to catch the whales and do this work. Obtaining oil was laborious and time-consuming — a 16-metre whale could take one or two weeks to process, and injuries from the sharp tools were frequent. And since the blubber was left on the beach to ripen for a week or two (to make it easier to cut up) and the gutted carcases were simply shoved back into the water to rot, the foul odour associated with the job was indescribable, and added to the stench of the boiled-out 'mince' being burned as extra fuel on the fires, and the raw oil bubbling in the try-pots. The special breed of workers gradually arrived. Undismayed or unknowing, Bosworth, Bundy and Wright somehow heard about the new undertaking and turned up, and other European men-of-all-trades drifted in — a mixture of adventurers, ship deserters, escaped convicts and absconders from other stations, hard-bitten and difficult to manage, but between them able to face the conditions and build and staff the station.

To obtain the necessary equipment, Barrett and Love had to link up with a patron in Sydney, and it could safely be assumed that it was around this time that the two of them contacted and became agents for Richard Jones. He was head of one of the chief merchant houses in Australia and had moved into whaling in 1829. With him as a consignee, over the next few years they gradually built up a makeshift team and the extensive apparatus needed to catch and process the whales.

The men who stayed soon linked up with local female partners. The hapu always provided long-term visitors with regular partners anyway, to prevent 'dalliance with the wives of others',[8] but Barrett and Love, as the whaling village headsmen, were responsible for arranging more formal 'marriage' contracts with the chiefs of the women's whanau. They made it clear to the men that these arrangements meant the 'wife' could be swapped for another if she turned out to be lazy; but that, in return for the women's sexual and housekeeping services, the men must treat the women and their families well and recompense them if they parted. The deal also involved the men in some tribal responsibilities. Part of the latter — defending their households against enemy attack — became apparent more quickly than the 'husbands' may have anticipated.

Local Ati Awa, keen to work for Tiki and Hakirau, also joined the team.[9] They either built their own whare on the station, or stayed with female relations living with European whalers — one way in which the latter honoured their 'marriage' contracts. They slept by the fireplaces and paid for their keep with gifts of potatoes and pigs. Both races began to learn a little of each other's language, but pidgin Maori, developed by whalers and traders around New Zealand, was the common tongue.

Gradually, the tutu, hebe, kotukutuku and flax in the bay gave way to a growing village of dwellings. Barrett, eventually, owned one of the few homes built of timber. His house, more substantial than he would ever have owned in England, stood on a rear knoll near the middle of the bay and had a fine view over the whole station and out to sea, and was far enough away from the works to avoid most of the stench. Guard's house, for summer use, was thereabouts too. Love settled his family in a cottage on the level ground at the foot of the knoll, and Keenan tucked his home into the hill on the northern side.

Most of the whalers' houses had thatched roofs, kareao (supplejack) walls filled with clay, and beaten earth floors; but there were differences. Ati Awa whare had low entrances, a hole in the roof to let out smoke, and woven mats on the floor. The European cottages had high doorways, chimneys and, often, more than one room, with bunks, pistols, muskets and lanterns lining the walls and the seamen's piles of ropes, oars, equipment and clothes covering the floors, along with barrels of provisions. It was the well-organized chaos learnt at sea that they taught their wives to maintain.

When they could afford it, Barrett and Love clothed their oarsmen of both races in European seamen's blue trousers and striped shirts. Rawinia and Mereruru were now for the first time living in a community where European influences were prominent and they, too, along with the whalers' women, adopted European dress, wearing loose, printed calico gowns, but usually with their blankets around their shoulders. The women as a whole kept themselves well-groomed, their houses clean and neat, and proved to be faithful and adaptable partners — loving to their men and to the children who started to arrive. They began to serve turnips, cabbage, ship's biscuit and bannocks for meals, alongside fish, wild pigeon, kumara and karaka berries, as the couples got used to each other. They also smoked pipes and regularly sat gossiping in their doorways along the village's interlacing tracks, just like any other housewives of the time. Barrett and Love, as headsmen, handled the inevitable sporadic domestic and village disputes, keeping the peace among their own people and ensuring good public relations with the local chiefs.

As he had done at Taranaki, Barrett introduced barter as a method of trade and the tribespeople began to grow produce for the whaling stations and passing ships, as well as for themselves; but he and Love adopted the normal method of remunerating the whalers. Each man received a 'lay', a ratio of the season's profits, gauged according to proficiency and skill. This was entered as a credit for each individual on the station's books; then Barrett and Love sold the whalers goods bought cheaply through Jones in Sydney — tobacco, clothing and other wants — at exorbitant prices. The men were thus kept in perpetual debt, and had to work for them the next year to pay it off.

The only item station managers ever sold at a low price was arrack rum, the cheapest and roughest on the market, made readily available to anaesthetize the men against the rigours and stresses of their occupation. Toms made his casks of rum go further than anyone: '. . . when I takes out a glass of rum, I puts in a glass of water; when it gets too strong of water, I puts in turps; and when it gets too strong of turps, I puts in bluestone.'[10] The ensuing constant violent quarrels and drunkenness were part of the whaling subculture of the times, and a function of growing profits. It was the women who maintained a fragile stability in the community, by exerting subtle leverage through their marriage contracts. More important features for the headsman at every station were their men's familiarity with, and adherence to, the code of whaling laws,[11] and that they were extremely well disciplined while working; and they all were, since their lives at sea depended on concentration and co-operation.

A small core of white men could endure the often dangerous and unpleasant conditions without the crutch of European alcohol and, in the early days, the tribesmen did not indulge in drunken revelries either, although they had their own brews, made from aged, or rotten, corn or potatoes. The tribesmen were cheerful and willing on the job, quite aware of what was due to them and had equality of pay and conditions in the workplace, but once off-duty they discarded their European clothing, wrapped themselves in blankets and, essentially, lived much as before.

A favourite meeting place after hours was Barrett's house. His natural management style, and Love's, was a combination of blunt good humour, firmness and kindness, in contrast to the often harsh and brutal methods of their contemporaries. Men of both races dropped in often to relax in front of the Barrett fire, to chat, smoke, carve scrimshaw, or tell exaggerated yarns in the fashion of the day. Barrett could match the best of them when it came to entertaining with factual or adorned stories.

Diversions aside, after the season opened at the beginning of May the focus for everyone was the right whale,[12] whose V-shaped misty blows were visible up to 18 kilometres away on a fine day. Each dawn, scouts sat on the hills opposite Te Awaiti and boats' crews rowed out, in anything short of a violent storm, to the entrance of Kura-te-Au to look for their quarry. Barrett and Love would have had the final say as to whether it was too risky but they would seldom have prevented the boats going out, as the men's income depended on their bringing in a catch.[13]

The whaling boats were clinker-built, sturdy, swift, slim and buoyant, ranging in length from 6 to 9 metres and proportionately wide. Each one was imbued with a personality, decorated with paint, and frequently given such nicknames as 'Swiftsure', 'Kangaroo' or 'Saucy Jack'.[14] Pointed at both ends for better manoeuvrability, higher at the head and stern than amidships, they carried five to seven oarsmen using oars of different lengths (because of the unequal numbers rowing), who greased the metal rowlocks or substituted others made of wood, rope or leather, in their efforts to row silently and catch the whales unawares.[15]

The harpooner rowed in the front until he took up his weapon to shoot, when, of course, equal numbers of oars would be left. A headsman commanded each crew and steered from the stern, using as rudder an oar 6 or more metres long. These men too had colourful nicknames, which sometimes guaranteed them necessary anonymity: 'Flash Bill', 'Gypsy

Smith', 'Black Peter'.[16] Neatly tucked on board were mast and lug-sails, hundreds of metres of rope coiled in tubs (sometimes made of flax by the men, since flax, because of its lightness, could be half as long again as hemp rope), several harpoons, knives and other spare equipment — all arranged for quick access — and water, grog and biscuits in case of a long pull home. Each boat represented a goodly investment of capital. Saving every penny, Barrett and Love must have built up their 'fleet' till they could send out the optimum hunting pack of six boats.

When they spotted their quarry, the whalers would all lean lustily onto the oars until they neared their target; then they paddled as silently as possible to get close. In calm to moderate seas this could be done safely with a measure of certainty. Right whales were then easy targets as they swam along the surface not much faster than 10 kph, their narrow open jaws capturing tiny plankton on their fine bristles. But in strong winds and gales, difficulties and danger were constant, and one blow from a sounding whale's tail fluke (possibly over 5 metres wide) could demolish a boat completely. Despite all precautions, the station's boats were often stove, gear lost or demolished, and men injured or killed.

Whatever the weather, when they were within striking distance the harpooner in the nearest boat flung the iron javelin into the whale's calf. Everyone waited during its ensuing death struggles until the mother returned to her dead baby, then the boat nearest her threw its javelin into her side, although by the laws of whaling, whoever harpooned the calf had the right to the mother. Once the adult mammal was struck, a potentially lethal game began for both whalers and whale.

The crew attached by harpoon quickly 'peaked' their oars and the accompanying boats rowed away to give them space. As the whale sounded, the ropes snaked out at a speed sufficient to burn flesh or whip a man overboard, and it became one man's task to throw water, to prevent scorching of wooden equipment. Then they all hung on grimly as the whale pulled them along at high speed. When the distressed beast finally surfaced for air, the other boats of the team closed in to fix new harpoons if necessary and, after much leaping and plunging and spout-

*Two boats, from the usual team of six, close in on a right whale, with a bay whaling ship in the background. The crew who harpooned the mammal were awarded their lay in decreasing amounts, from headsman down.* AUCKLAND PUBLIC LIBRARY

ing of blood, the headsman finished off the victim by throwing one, two or three of his bright, oval-pointed lances into its threshing body.

There were many days when the boats returned unencumbered, but with a kill there were usually long and arduous kilometres to row towing the carcase, tail first. The rowers took it in turns to have fifteen minute breaks, until they reached base. Sometimes, when the weather was deteriorating, the unwritten laws of the sea had boats from their own and other Te Awaiti stations come out to help. Then, like several odd-legged spiders dragging their prey into their tunnel, the boatloads of men strained to pull the enormous black dead-weight through the channel entrance and down to their bay. Even that close to home, they were not safe. There was a dangerous rip where the channel waters met Cook Strait and strong spring tides created mayhem trying to push vast amounts of water through the narrow channel opening.

Sometimes, when a kill occurred just before dusk, the boat would be too far from home to return safely; then a flag was stuck into the carcase to show ownership. All harpoons carried a private mark, but most of these flagged bodies either sank, or drifted away, or were eaten by sharks, birds or marine organisms. Then the crew would take shelter in a bay on the coast. Barrett's Inlet, and another known as Jackson's Boat Harbour, provided sparse sanctuary for the night or in dangerous weather, but there were only shellfish and penguins for food if whalers were marooned for more than a day. Love may have commanded boats at times, and Barrett was also known to have been a brave headsman,[17] but he tended to stay on land now as manager of the station store — and his girth must have increased along with the profits.

When the whales swam away, Barrett and Love had to see that the maintenance work, on-going all season, was completed, before the men went off-duty. The station hands had to mend the boats or build new ones from the island's timber, forge tools, harpoons and other damaged implements back into shape, or make new ones, weave and strengthen ropes, and see to a whole raft of other tasks. Only when these jobs had been attended to did Te Awaiti become all but deserted, usually from November to April.

During these summer months the families moved into the nearby villages, or into their own small houses in the hills, where they kept poultry and grew vegetables and other crops for themselves, and supplemented their lay by selling these to passing ships. Insidiously, the profits from trading increased the time the tribespeople spent on gardening, and reduced the motivation for fighting.

Barrett's well-known good humour on the job masked tremendous effort, of course. Numerous other factors had to be taken into account in this remote part of New Zealand: fluctuations in whale migratory patterns, availability of employees, provisions and whaling equipment, motivation of the teams, and competition with other whaling stations. When boats were damaged or destroyed, it was the two partners' responsibility to repair or replace them. Their principals in Sydney contributed nothing in this respect. Shipping from Australia, to carry their whale products, could be unavailable for long stretches, or lost at sea with a full load. Carpenters and blacksmiths were so scarce that they commanded exorbitant wages of ten shillings a day. And when workers absconded, the two friends were left with their bad debts.

Their operation was only one of several affiliated with R. Jones & Co. at Queen Charlotte Sound and elsewhere in New Zealand. Jones paid them only £10 per tun for oil, and sold it in England for three times as much, and £6 per ton for bone, for which he received £15. There was little they could do to extract more from him. His payment was the going rate and the oil profits flowed not to them, but into Jones's coffers via the *Vittoria*, *Eleanor* and the *Success*, the last captained now by Buckle, Barrett's old acquaintance from Ngamotu days.

In times of record oil tonnage, the smog emitted by try-pots in the whale fields became like 'London Town',[18] as these and other vessels ploughed back and forth across the Tasman as quickly as their owners could turn them round. In the first three years of Barrett and Love's operation, tuns reported as 'brought up' to Sydney from five New Zealand stations, including theirs, showed more than substantial increases: from 409 tuns in 1833, 849 in 1834, to 1231 in 1835.

Despite the dangers and low profit margins, however, Barrett and Love stayed in the business, and earned enough to build 'summer residences' for their own use (probably at Anaho, towards the northernmost point of Te Wai Pounamu) and to build up their fleet and equipment to an impressive extent. Their families grew too. A third Barrett daughter, Hara (Sarah), was born in 1835, to join Kararaina and Mereana, while the Love's fourth child arrived at about the same time.

Events were now combining to push Barrett into a more prominent role in the country. Since he had left Britain, three companies had been formed to promote the colonisation of New Zealand. Each one had arisen from the ashes of its failed predecessor, and their core members had been inexorably pressing the Colonial Secretary to help them to settle those southern islands near Australia. Sydney merchants and Church Missionary Society (CMS) representatives had contributed, too. The former had played their part by petitioning His Majesty's Government to establish a military presence in New Zealand, overtly to control drunken and violent men, who had adverse effects on tangata whenua, but covertly, to protect their own mercantile enterprises and to discourage American and French competition. The latter, in the form of British missioner Henry Williams and other colleagues, had supplied their input by persuading Ngapuhi chiefs to request the Crown's protection at about the same time and for the same reasons.

Barrett may have been aware of rumours about these events, along with political snippets from New South Wales, missionary news, whale statistics and the general information that visiting seamen brought in as grist for the station's consumption. Tribespeople crossing Raukawa Moana in canoes were other news-carriers, through whom Te Ati Awa remained *au fait* with one another's doings — and with the activities of other iwi. The Barretts would have heard that Waikato made the expected retaliatory attack against Ngamotu at the end of 1833, and that its villagers had taken refuge at Mikotahi. A stalemate siege had developed which resulted in Rawinia's mother, Poharama, and several other Ati Awa chiefs being taken back as vassals to Waikato as part of the peace process. After that, a few Ati Awa stayed on at Moturoa, eking out an existence and maintaining a hold on their hereditary land;[19] but most of the others melted away from Ngamotu, down through Waikanae to Whanganui-a-Tara and Te Wai Pounamu. Some had benefited when Pomare, principal chief of Ngati Mutunga, persuaded that iwi to emigrate from Whanganui-a-Tara to Wharekauri and gifted Ngati Mutunga lands at Whanganui-a-Tara to Ati Awa and Taranaki hapu before he departed.[20]

Whatever Barrett and Love made of the general tidings, in 1834 or 1835 the seeds of Te Awaiti station's decay had been unwittingly sown. At that point, a trio of American ships joined the seven other European bay whalers in Cloudy Bay. They provided stimulating contact, but caused incalculable harm. As a result of their enthusiastic reports on their return home, swarms of profit-hungry American vessels sailed down to New Zealand over the next two seasons, and competition for the whales became patently uneconomic in the Raukawa Moana killing fields.

That the 1836 whaling season was going to be difficult must have become apparent to everyone involved, once they counted the numbers on the water. As one captain put it, '. . . the sight of seventy or eighty boats bearing down caused whales to turn tail and escape . . .'.[21]

Not only did these unwelcome visitors anchored in Cook Strait have access to the whales before they came within reach of the shore stations, but they took females for preference, leaving the shore whalers the bulls, which produced less oil. Over a six-week period in Barrett and Love's area, only thirty-one whales were taken by twenty-one of the American ships. Whaling became quite uneconomic for all the bay whalers, and disastrous for Barrett and Love and the other shore stations. Cloudy Bay had its European population at last, but it was in vessels, not in towns, and it was wiping out the source of income it was pursuing.

At the end of the season, lays were almost non-existent, and Heberley, who had left Guard's Kakapo station for Te Awaiti in September 1836, found his new employers making no concessions. The difficulties in New Zealand were exacerbated by an Australian seamen's strike for higher wages, which erupted two months into 1837. Jones and his wealthy Sydney colleagues broke the strike, but it affected Barrett by making the availability of shipping across the Tasman even more uncertain. Everyone needed every penny they could earn. During one or two summers, Barrett took the *Harriett*, a 42-tonne cutter, belonging to Toms, and went hunting seals down the west coast of Te Wai Pounamu and coal at Cape Farewell.

Astonishingly, Te Awaiti was attacked again, in the heat of February 1838, by a group of usually friendly Ngati Toa, led by Te Rauparaha himself. 'Old Robulla' — an oft-used seamen's nickname for Te Rauparaha — had gathered a band of warriors and started south to confront the long-planned Ngai Tahu taua, which was travelling north to eject him from Te Wai Pounamu. Two events changed his plans. Firstly, epidemics of measles, influenza and tuberculosis put the members of the travelling taua out of action.[22] Secondly, Te Rauparaha discovered that his Ati Awa allies, travelling ahead of him, had eaten wild turnips growing on a Ngati Toa burial ground. Allies or not, this breach of tribal etiquette could not go unpunished. The main enemy had to wait (he was unaware that the taua had dissolved because of illness). His men caught up to and attacked the offenders, stripped them naked, and removed their canoes, weapons and all possessions in retribution. Not surprisingly, Ati Awa, also acting under tribal convention, had to repay this indignity. Once clothed and armed again, they chased after Ngati Toa, who had meanwhile withdrawn as far as Te Awaiti. The two sides met and conflict began, only 300 metres from the station's store. Other accounts give different reasons for the strife, but whatever they were, the uninvolved Europeans at Te Awaiti hastily took shelter on ships lying handily offshore.[23]

Both station managers, with Heberley, were absent in Australia at the time. In December 1837 they had quickly responded to a plea — and promise of reward — from a Captain Hay to sail his ship, the *Hannah*, back to Sydney to replace its crew, which had deserted at Totaranui. Hay had caught them at a low ebb. Whaling elsewhere in New Zealand was still viable, but returns at Te Awaiti had been just as low in 1837 as in 1836, while station overheads remained substantial.

At this point, Barrett and Love may have intended to walk away from their holding and cut their losses, since they surprisingly took their wives and children with them to Australia. Heberley left his wife at home. Once in Sydney, however, they heard at 'the office' that '. . . the Southern part of the North Island was going to be settled by the English . . .'[24] but from England, not Hobart, as suggested earlier. This may have changed their minds, because the three men bought blankets and tobacco with the £400 Hay paid them, and returned to Te Awaiti determined to buy well-situated land and stay afloat until their new hopes for an English settlement were realized.

When the *Hannah* returned from Australia and dropped anchor off the station during the fighting, the two families and Heberley were spectators as bullets flew, the station cannon roared and weapons clashed.[25] Only when darkness fell could they land. Apart from the loss of 'a quantity of potatoes',[26] no property had been destroyed, but a few more tribal graves had to be dug. Utu had apparently been achieved on both sides and peace was made that night; but, just to make sure that 'his pakeha' remained safe, Te Rauparaha asked if they would 'kindly . . . send a bottle of brandy'. Nicely fortified, Te Rauparaha stayed about three days without troubling them further, then continued southwards.[27]

There was little cause for optimism in the months ahead. The 1838 whaling season was a 'total failure' and Love became chronically ill. Barrett struggled through the 1838–39 summer months, keeping the *Hannah* sailing back and forth, filled partly with Te Awaiti whalebone and dried fish for his principals in Sydney, and partly with small whalebone consignments for other Australian merchants from various stations further south. There was just enough trade to warrant *Hannah*'s employment, despite the fall-off in whale products. The 1839 season produced very few catches of smaller whales. Starvation was never a prospect, but the zest for life had faded with the whales, and Barrett's outlook was dismal. Love was failing and Rawinia and Dicky took his and Mereruru's eldest son, Hone Tanerau (Daniel), into their home. An air of gloom hung heavily over the station.

However, events had occurred which Barrett may have noted with optimism. In September 1838, HMS *Pelorus* had anchored off Te Awaiti, while cruising the area to do what the British Government had been only talking about for years — make sure that English people in New Zealand were behaving themselves and were safe. And on 20 June 1839, John Bumby and John Hobbs called in at the bay, having put some land under tapu for a Wesleyan mission at Whanganui-a-Tara. The rumours of English settlement now took on slightly more substance.

*Chapter 4*

28 August–25 September 1839
BUYING WELLINGTON

---

It was in the disquieting atmosphere of uncertainty at Te Awaiti that the whalers, on 28 August 1839, saw a strange European and a tribesman walk into the village from the bush track on the north side of the bay. The white man spoke English, introducing himself as Doddrey and his companion as his guide, and said that a Colonel Wakefield had sent him to find their community. The colonel was at Anaho, he explained, on a British ship, the *Tory*, carrying officials of an English company that was going to settle New Zealand. Suddenly, the future was there.

Barrett did not respond by going himself to meet the vessel, since Jacky Love was obviously now dying.[1] Also, the station was short-handed, with tribal whalers having left to attend the tangi of Te Rauparaha's sister, Waitohi, who had died the previous week at Kapiti. They had departed fully armed, since rumours had come of further friction there. Although the season had been so poor that the lack of workers was not crucial, it was unsettling; and the boats still had to be sent to sea. Barrett remained close to Love, and the next day despatched two of his senior men, Elmslie and Williams, with a whaleboat and crew, to return Doddrey to the *Tory* and to pilot the ship to Te Awaiti.

A day and a half of suspense and dreadful weather followed, until a nuggetty barque appeared around a headland, being towed against the tide by her sailors in row-boats. The *Tory* was around 400 tonnes and sported eight guns and an unusual male figurehead on her bowsprit, which turned out to represent the Duke of Wellington. Te Awaiti was doubly lit up for the ship's arrival: large fires under the try-pots glared through the gathering darkness, and the killing of a large whale that morning was being celebrated with great gusto and 'profuse libations'.[2]

*Tory Channel and Te Awaiti, September 1839, sketched by Charles Heaphy. The giant sheerlegs can be seen at the right-hand entrance to Te Awaiti, the lookout post on the hills opposite, and the distance to the entrance of the channel that the whalers had to row at dawn each morning during the season, before they went after whales.* ALEXANDER TURNBULL LIBRARY

The rowers rested on their oars and the *Tory* dropped anchor off Te Awaiti at sunset. It was Saturday 31 August 1839.

Barrett had himself rowed out to the ship and was greeted by a group of men, including Williams and Elmslie and a well-groomed man, slightly older than Barrett, who introduced himself as Colonel William Wakefield. Among the others were Dr Ernst Dieffenbach, the expedition's naturalist, Dr John Dorset, to be the colony's surgeon — both aged about thirty — Charles Heaphy, artist and draughtsman, and Edward Jerningham Wakefield, 'secretary' to the colonel, his uncle, and gentleman-at-large, both nineteen, and Edmund M. Chaffers RN, the *Tory*'s captain.

The total number on board was thirty-five and all spoke of being very impressed by their first contacts with New Zealanders. They had been told that the local people were sinister and menacing, but had found them open, confident, hospitable and helpful. The ship carried fresh potatoes, pigs, pigeons and parrots, provided by people at the *Tory*'s previous anchorage at Anaho, and everyone was luxuriating in clean clothes, laundered for them by the womenfolk there. No, they needed nothing from Barrett's store.

Wakefield ushered everyone below decks, and began plying Barrett with questions and grog. With Williams having an immense girth and Barrett being as round as he was tall, their extra bulk, together with Elmslie, may have made the small 'great' cabin uncomfortably crowded, as they all settled down for the evening's discourse. In England, Barrett and his colleagues would never have expected to sit down and take sherry with a colonel, nor would they ordinarily have socialized with the other expedition members he had just introduced; but now they settled right in and began entertaining the *Tory's* travellers with extravagant tales of their adventures and reminiscences from the past decade.

William Wakefield was thirty-six, of medium height, slim, with an aquiline nose, upright stance and brown hair shaped fashionably to his head. He was smartly turned out — he travelled with a valet — but his superficially attractive and sympathetic personality had an extremely reserved and complex side, which eventually baffled everyone who knew him. He had left a thirteen-year-old daughter in England, and was already a widower.

After the initial good-fellowship, he injected a sense of urgency into the conversation, explaining he had just two months in which to buy land for hundreds of English settlers who were already under sail for New Zealand. By the end of an evening of alcohol, smoke and animated talk, a slightly befuddled understanding had emerged. Wakefield reported to his directors that Barrett said he 'would not mind coming over with me',[3] Jerningham Wakefield wrote that 'it was determined that Barrett should explain our views to them' (the chiefs),[4] while Barrett's memory was that he was offered the jobs of piloting the *Tory* into Port Nicholson, then acting as interpreter while Wakefield bought the land.[5]

It is therefore unclear whether Wakefield asked Barrett to join him, or whether Barrett seized the opportunity to insinuate himself into the colonel's team, with an eye to launching himself into a new career or, perhaps, to bring benefits to Rawinia's hapu, by introducing this Englishman to them first, to buy their land; but, all unknowing, he had teamed up with a man of complicated character, given a very difficult

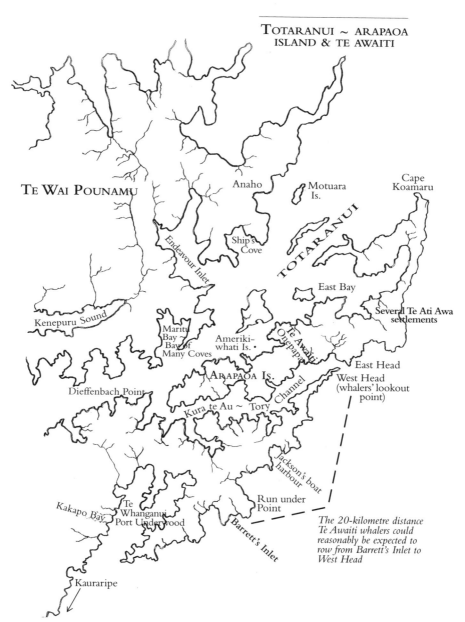

*Arapaoa Island, Te Awaiti and environs, showing the extent of Te Awaiti whalers' hunting grounds. It was only a two- or three-hour walk across the hills for the Maori from Te Awaiti to where William Wakefield and Barrett negotiated the East Bay Deed with them in November 1839.*
ADAPTED FROM AUTOMOBILE ASSOCIATION MAP OF MARLBOROUGH SOUNDS, 1986.

*Edward Jerningham Wakefield.* ALEXANDER TURNBULL LIBRARY

brief with an impossible time constraint, as agent for a company defying the British Government (see Prologue).

An immediate understanding apparently sprang up between Barrett and the Wakefields. Jerningham reports how everyone on board warmed to Barrett's 'jovial, ruddy face, twinkling eyes and good-humoured smile', while listening to the 'wild adventures and hairbreadth scapes' with which Barrett and his two companions entertained the cabin gathering.[6] Wakefield summed up the whaler as 'a respectable man, desirous to give me his valuable assistance with the natives of Port Nicholson', and added him to his team.[7] He thus supplanted Ngaiti, a young 'chieftain', smartly dressed in English clothing, who had been brought from England as interpreter and who would have been listening to all this.[8]

To Barrett, Wakefield appeared as a gift from the Almighty, to whom Barrett was wont to show the usual respect. Not only would his old friend, Love, die soon, but their business was on the rocks and Jones in Sydney was having financial problems.[9] Wakefield had told him that he represented a so-called New Zealand Company, which aimed to bring out to New Zealand hundreds of English families — probably to Whanganui-a-Tara, that wonderful harbour which Wakefield knew as Port Nicholson — to settle and to create the 'commercial Capital of New Zealand'.[10]

Barrett may have imagined land values at Whanganui-a-Tara soaring. Te Ati Awa had been bequeathed land there by Ngati Mutunga when the latter moved to the Chatham Islands. Rawinia's family may really profit. If he played his cards right, he would be able to leave whaling, and move on from being mate to Love in a dwindling trading enterprise, to becoming the colonel's right-hand man. Of course, New Zealand should be settled by British people and become part of the Empire; but, with his help, Te Ati Awa would be the main beneficiaries and, with the colonel's help he, Dicky Barrett, could make a fortune.

Later, when the maps were brought out and subdivisions discussed, Barrett may have had doubts. He would have seen that the parallelograms on the map bore no relationship at all to the physical contours of Whanganui-a-Tara (see page 142). Someone with infinite ignorance had drawn plans of streets and sections, parks and cemeteries, in an area which, he knew, consisted of swampy delta or high hills and steep gullies. Wakefield had instructions to buy enough land in this impossible terrain to accommodate 1,000 families, so that they could have one town acre (just under a half hectare) each and 100 country acres (about 40 hectares) not far away, on which they could grow their food and support themselves initially. It seemed quite impractical; but the immigrants had already paid their money and were on their way, hence the rush.

On nearly all counts, the rest of Wakefield's instructions seemed totally overwhelming.[11] It is worth detailing some of the main points contained in the several closely written pages which Wakefield must have paraphrased for Barrett's briefing. He was to buy all the best areas of land in the whole of New Zealand, to make it unavailable to others. More important than fertility of soil was that the areas had natural features such as good harbours, wide rivers and waterfalls to facilitate communication and transport. These sites were to be chosen with an eye to the development of the central commercial capital of New Zealand as well as secondary towns. Wakefield had to make sure he bought land around one good harbour at least on each side of Cook's (*sic*) Strait, close to good agricultural or commercial land, and Port Nicholson was mentioned as a possibility, since it sat on the shortest route from Australia to England. This was where Barrett could be useful.

Wakefield had two months to finalize his purchase of land in this region, then he was to travel north and repeat the process there. Kaipara was mentioned as a starting point — where the New Zealand Company had some land already that he was to acquire officially — but he had to make sure there was not a better place for the establishment of the seat of the commercial capital in northern New Zealand. At all times, he was enjoined to treat the 'natives' with openness. They were suspicious because of the ill-treatment they had received previously from Europeans, therefore he had to make sure that all chiefs and tribespeople understood and approved completely of the purchases he wished to make, and that they all shared in the purchase goods. They also had to know about the large numbers of people who would be arriving; and that one-tenth of the land would be kept for them as special reserves, on which they would live among Europeans. This detail quite ignored the fact that, as Barrett knew, whanau preferred to live close to one another in a village community. On top of this, he had to explain that, though they might think themselves disadvantaged and inferior without their former properties, this would not be the case. The reserves would increase greatly in value because of the European impact. For each purchase negotiated, boundaries had to be walked, the area stated in words and a plan attached to each contract; and Wakefield had to bargain shrewdly so that he paid only the 'sufficient price' without it being 'less inadequate' than what had generally been paid to New Zealanders in past land deals.

So far, the information for Barrett was straightforward; but there was more, which the colonel would not have bothered to include at this stage. He had to make sure that he bought for the New Zealand Company in places where the 'natives were keen not only for English settlers but for English wages. Wakefield was to prepare them for this employment ethic by paying them 'liberal wages' to fell timber and take logs to earmarked town sites, ready for building, and to collect and prepare flax and spars for return freight in vessels bringing out settlers.

And all New Zealand Company employees had to be monitored carefully. They were to be dismissed if they engaged in rude or unseemly behaviour towards the natives, of if they were drunk in public. Employees were to show deference to missionaries and talk about them with respect to the natives, to facilitate dealing with the latter, who had a high regard for the missionaries in the north; and in the same vein, there was to be no work on Sundays, but assemblies for worship, because to the natives such piety indicated that the Europeans were rangatira or gentlemen. Ngaiti was to be cosseted and held up as an example.

Enclosed with the instructions was a letter from the New Zealand Company Secretary, offering guarded advice as to how Wakefield may more easily achieve the company's purpose.

\* \* \*

Although Barrett had heard the rumour about, and eagerly looked forward to, the English coming, he could never have expected to be one of the facilitators for such an expedition. As the evening progressed through the 'getting to know you' phase, and into discussion of the details of the New Zealand Company's plans, Barrett could have been excused for some doubts, as he absorbed the implication of the future. However, his equanimity evidently undisturbed, he did not disillusion them of their belief that they could bring a large body of settlers to live in the Whanganui-a-Tara district without too much trouble. From Elmslie's description of how the whalers and others fenced the land and ran stock on it without the locals minding, Wakefield had gained the impression that his large body of settlers might do the same. Intentionally or not, Barrett added to this misconception by intimating that unoccupied 'waste' land was not 'thought of or claimed by anyone'.[12] The whaler should have known by then that there was no such thing as 'waste' land in New Zealand. Added to that, with some circumlocution, he gave the colonel and his friends the impression that their 'wish to purchase a large district of waste land would be looked upon as a novelty' by the tribes.[13] Indeed, he suggested he might be able to buy Port Nicholson for the New Zealand Company, because his wife's family belonged to the tribe who lived there.[14]

Since Wakefield seemed to have his eyes on Port Nicholson, Barrett attempted a succinct, if slightly confused account of the recent changes in ownership around the harbour, and the complicated tribal claims to it.[15] He made no secret of Te Rauparaha's tribe's dominance in that area, but he was unaware that another Ngai Tahu taua, all fired-up to reassert their ownership, was about to make another effort to dislodge Te Rauparaha and Ngati Toa from Te Wai Pounamu. He did intimate, however, that Ngatiawa (*sic*) seemed able to supplant Ngati Toa only because of their 'very superior numbers'.[16] His thoughts may already have been turning towards Te Puni and Te Wharepouri who, anyone would have acknowledged, had a not inconsiderable claim to the area of land Wakefield seemed to have his eye on.

In the warmth of the night's farewells, the colonel said he intended to name the channel his men had rowed up, Tory Channel, and Barrett invited Wakefield and his team up to his home the next day.

He met them on the beach, when they rowed in after morning prayers. His house stood on a middle knoll, at the back of the settlement, outlined against a backdrop of dark-green bush profusely draped in patches of clematis; but Sunday or not, the trying out of the precious whale continued, and the English gentlemen were forced to make their way to it past the tryworks in full glowing and smoking stench, over the bloody and grease-clotted sand covered in skulls, vertebrae and bits of decaying whales, and through the busy whaling village.

Once inside Barrett's home, they were in another world. The dwelling was strongly built of sawn timber, with wooden floors and walls, and a wide verandah. Rawinia, a tall, dignified woman now in her late twenties, greeted them, and seated them by a glowing fire in the main room, round which several tribespeople and whalers were making themselves at home. The Barrett daughters, Caroline, Mary Ann and Sarah, aged about nine, seven and four, plus Daniel Love, the same age as Caroline, were part of the group, all cheerful and full of shy chatter in Maori. Rawinia would have been one of the reasons that Wakefield reported to his directors, '. . . many wives are also entitled to every praise for their fidelity, care of their children and industry ... and are fit to take a very respectable station amongst English matrons....'[17]

Barrett and Wakefield continued their discussions begun on the ship. Wakefield referred to Elmslie's conversation about white men clearing and cultivating land without tribal interruption, and this time Barrett disillusioned him. The lands were not all 'waste'. The tribes would need good payment, he stated firmly, especially if their villages and cultivations were in the areas that the Englishmen wished to obtain. Barrett's remarks, sprinkled throughout Jerningham Wakefield's later writings, are not constant. Either Jerningham was inaccurate, or Barrett vacillated. Certainly, Jerningham was left with the impression, from Dicky's earlier tales of tribal battles and conquests, that '*might* constituted the only right', as regards ownership of villages and cultivations.[18]

Later, Barrett and Heberley showed the visitors around the village. There were about forty whalers' cottages, much simpler constructions than Barrett's, built of reeds and rushes, usually with two holes furnished with shutters for windows. Sheds and equipment scattered round, a dozen whaling boats lying on the shore — nine of which belonged to Love and Barrett — and gigantic sheerlegs towering over them, displayed the two whaling partners' achievements over the past six years.

Women, sitting at the doors of cottages they passed, took the pipes from their mouths and greeted the newcomers, 'Tenei ra kokoe pakea', as Jerningham wrote. The Europeans had to thread their way through a score or so of their dark- or blue-eyed, glossy-black or fair-haired inquisitive children as they explored Te Awaiti and then traipsed over the hill to meet Jackson at his small whaling station in the next bay.

Everyone they met looked expectant or quizzical, obviously sensing the possible benefits to themselves of Wakefield's project. Ati Awa were readying seed to produce 'larger quantities of potatoes than usual' for the expected immigrants, while the Europeans talked of raising greater numbers of pigs and poultry.[19] One James Wynen was ahead of the field, sent by his principals in Sydney to invest in land in 'the best situation',[20] before the rumoured colonists arrived.

Later, Wakefield explained he wanted to explore the Sounds for promising land and anchorages, as requested by his principals, and Chaffers thought he might search there for new spars. Barrett, with just two weeks left of the 1839 whale run, Jacky to care for and arrangements to make for the station before he left with Wakefield, could not be absent for this purpose. Jackson, not disinterestedly, suggested John Guard show Wakefield around; and, because of the bush telegraph, Guard and Wynen materialized on 6 September with a strong sailing-boat, crew, 'provisions for a few days' and blankets, and took Wakefield and his nephew to explore the whole area. Jackson followed a day or so later, probably hoping to take a punt on any land he saw Wakefield interested in.

While Barrett was organising the station for his coming absence and briefing a manager, word came that Henry Williams, a CMS missionary from the Bay of Islands, had sent a missionary schooner to Port Nicholson, to tell the inhabitants to hold onto their land. This could be another complication in Wakefield's negotiations for Whanganui-a-Tara, Barrett may have told his new employer, coming as it did on top of the visit three months earlier by Bumby and Hobbs, who had also sought a foothold for their church there.

Hearing about Williams's message on his return to Te Awaiti on Monday 16 September, Wakefield reacted quickly. Instead of buying land initially around Wairau and Hoiere, as Guard and Wynen had hoped, Port Nicholson became his immediate target. A slight earth tremor gave Wakefield a portent of what might be ahead as, in haste, the space between decks on the *Tory* was cleared and Heberley, the Barretts and several of their extended family moved into the quarters created.[20] The wind, however, took no notice of Wakefield's wishes, and they lay exasperatingly becalmed for two days until, on Friday 20 September, the breeze freshened. They weighed anchor at daylight and left Te Awaiti behind 'with a fair wind and tide'.[21]

It took them to the middle of that afternoon to reach the entrance of Whanganui-a-Tara. High rugged hills had faced them as they sailed across the strait. As Chaffers nosed the *Tory* into the harbour mouth, these sloped down, only to curve round and rear up again to port and starboard, forming a dark-green, jagged frame around the wide-spread circle of water before them. Although Barrett, in his role as pilot, had to alert the helmsman to a dangerous reef of low rocks off the headland on their left, it was a magnificent haven — like an inland lake. The hills lay high and deep in serried ranks, backed by snowy ranges. Wakefield noted them as 'by no means of the formidable height of those in the Sound . . . and present no obstacles to their cultivation';[22] but Barrett saw it as an impossible place for the planned town and country sections proposed by Wakefield. The only level land available for buildings, as Barrett pointed out to the colonel, was about 10 kilometres away at the far end of the harbour. Here, the Heretaunga River ran into the sea, creating a 4 kilometre-wide delta of swamp and sands. Other flat terrain was occupied by various small kainga. They appeared to be insignificant; but their occupant chiefs had rights to all the land around them.

The *Tory* had not advanced far when two canoes approached, paddled alongside and made fast. Te Puni with a few compatriots boarded from one, Te Wharepouri from the other, in a repeat of their encounter with the *Adventure*'s arrival at Ngamotu nearly twelve years before. Their appearance seems too much of a coincidence not to believe that Barrett had sent them a message in the interim. He introduced them casually as principal chiefs of this area and his in-laws. The colonel was impressed, but rather jumped the gun by suggesting that Barrett tell the chiefs that he would give them one week to think about selling their land!

The chiefs dismissed their canoes and sailed up the harbour on the ship, which anchored for the night about a kilometre or so from the beach in the shelter of Matiu, one of the islands at the far end of the harbour. Barrett slipped out of his pilot's role into that of interpreter and they all got down to business. He told the chiefs that he came to buy land in their country for the colonel and the New Zealand Company, specifically at Whangaui-a-Tara. They said '. . . they'd not deny me it because I brought their grandchildren'.[23] Te Puni showed real pleasure when informed that they wished to buy the place and bring white men to it and

Te Wharepouri reacted similarly and mentioned how the land had been made over to him five years before by Ngati Mutunga '. . . and there is no one who disputes his claim . . .', wrote Wakefield, a willing believer.[24]

Hundreds would! The present inhabitants at Whanganui-a-Tara, who had ousted Ngati Ira and other tangata whenua only over the past decade, were people who had lived uncomfortably side-by-side for centuries in the Taranaki district, and brought their prickly relationship to Whanganui-a Tara with them. Although their relationships were as mixed by inter-marriage as the smoke which arose from their fires, their disagreements about their tribal boundaries were similar to any family argument over property, exacerbated by the manner in which Pomare and other chiefs of Ngati Mutunga had disposed of their Whanganui-a-Tara land when they emigrated to Wharekauri in 1834. One account had Te Matangi and his son Te Manihera Te Toru, of Te Ati Awa, as the recipients, because they were related to, and had lived with, Ngati Mutunga since 1832. When this gift was reputed to have been made, Te Puni and Te Wharepouri, along with their hapu, had been still homeless after heke Tama-te-uaua, and it had suited them well when the two beneficiaries invited them to share the land. They acquired turangawaewae, and Te Matangi and Te Manihera gained the extra security that those two chiefs' warriors brought. The same account said that just before they left for Wharekauri, Pomare had given by panui the remainder of Ngati Mutunga's land to two hapu of the Taranaki tribe, who had settled at Te Aro. However, another scenario had Wi Tako's father, who was Ati Awa with Taranaki connections, inviting these two hapu to settle at Te Aro, after the previously mentioned fraternal in-fighting at Haowhenua.[25]

Whatever went before, a description of the main chiefs at Whanganui-a-Tara when the *Tory* anchored shows Puakawa and Taringakuri at Waiwhetu pa, Te Puni at Pito-one pa, Te Wharepouri at Ngauranga, Te Ropiha Moturoa and Wairarapa at Pipitea, and Wi Tako and his father, Ngatata-i-te-rangi, at Kumutoto. They were all Ati Awa and of almost equal rank. Ngati Tama, who were allies of Te Ati Awa, had lived at Kaiwharawhara since 1825. Tiakiwai pa was occupied mainly by individuals and groups of different iwi as a resting place on the track to and from Porirua. Taranaki and Ngati Haumia, the latter a hapu of Ngati Ruanui iwi, divided Te Aro pa between them. Some of these hapu would have claimed that Te Rauparaha still had a lien over Whanganui-a-Tara, while others would have argued that he had given it to Ngati Mutunga, because his niece had married Pomare. Amid the plethora of claim and counter-claim Ngati Ira, and other earlier tangata whenua, were living and fuming just over the hills, awaiting their chance to return.

Te Wharepouri and Te Puni were good friends, but they irritated other chiefs, and even their own brethren, by their high-handedness; and these to-ings and fro-ings had sparked irascible encounters. One such had arisen when chiefs of two Taranaki hapu at Te Aro had admitted wrongdoing to Te Wharepouri after one retaliatory raid. Penitently, in retribution, they had ceded him the Ngauranga land left them by Pomare; but, despite their making amends, he had destroyed one of their precious canoes. Aroused at this display of bad-temper, they cursed him and Te Puni and wanted nothing more to do with them.

Few Europeans at the time could have understood the complicated tribal political situations Wakefield was moving into. Rawinia may have been able to follow them, and their implications for her tribe, but it is unlikely that Barrett could have grasped them, or realized how tenuous was the claim Te Puni and Te Wharepouri made to rights over the whole of Whanganui-a-Tara. He probably made an initial approach to them knowing they controlled a few of the settlements, and hoping this would supply the land Wakefield desired.

Te Puni and Te Wharepouri gave no indication of these tensions to the colonel. With dignity and impeccable manners they remained to be wined and dined and to sleep on board. Wakefield listened with interest as, with Barrett interpreting, they gave him their version of land rights at Whanganui-a-Tara. In revenge for being cursed, they stressed how inferior to themselves the inhabitants of Te Aro pa were, describing them as taurekareka — a standard term of abuse meaning either scoundrels or, more disparagingly, slaves, as captives taken in war were called.

Jerningham picked up the latter meaning, thereafter referring to Te Aro as 'the slave settlement'. Thus developed a misconception which skewed the Europeans' understanding. The colonel would not have realized Te Puni and Te Wharepouri had two reasons to want him on their side — their wish to retaliate against the Taranaki chiefs and their concern at the presence of Ati Awa enemies, Ngati Raukawa, on the Kapiti coast.

It was a long session for these men on the *Tory*, and Barrett had to decide whether to translate all the comments and asides, or to select the ones which he saw as pertinent, or to translate only when asked. And should he confer with Ngaiti about subtle meanings or vital points, for the benefit of Te Puni's and Te Wharepouri's and the Europeans' understanding? While Wakefield explained one issue and Barrett interpreted, there was little check on accuracy or understanding since, by custom, the chiefs listened impassively, not seeking clarification in the middle of a translation.

Young Jerningham found the understanding between the two races to be quite satisfactory, but he was biased and hardly qualified to judge. There was tangible goodwill, but the insidious mixture of misunderstandings, which would dog Barrett's future, was beginning to accumulate.

Wakefield may not have realized that his summing-up of Te Puni and Te Wharepouri was a two-way process. He was giving them an opportunity to mull over possibilities, presented to them by meeting him first, before their rivals. They may have seen this as a unique chance to take complete control in Whanganui-a-Tara and to challenge Te Rauparaha's supremacy in the south of Te Ika a Maui. Barrett did not have the knowledge to contradict the two of them, nor was it his duty as an interpreter; but their campaign to brainwash Wakefield about their pre-eminence in the area badly needed a balancing influence.

As they ate and talked in the panelled cabin on their first evening in Port Nicholson, Barrett's attempts to explain Wakefield's concepts and objectives to the visitors must have been limited by the fact that, however willing, he came to the job ill prepared. His grasp of the Maori language, after about eleven years in the country, was adequate for his needs, but not for the ambiguous statements and power plays of the people among whom he would be working, their different attitudes towards the occupation of land, and the nuances of meaning in languages. Barrett had to try to explain 'buy' and 'sell' to the chiefs. They had no concepts to cover these terms, nor could they imagine the total alienation of land that such transactions involved. It seems that Te Wharepouri, for one, gained a glimmering of what Wakefield wanted; but his understanding did not encompass the hundreds of settlers who would appear, nor the long-term heartbreak and difficulties they would provoke.

While Barrett translated and struggled to explain the system of reserves to the two chiefs — in a way he himself later confessed was minimal and vague — he had not realized that Te Puni was scarcely listening. At this stage, any plans or payment that Wakefield might offer were important but secondary. Te Puni later admitted that he was daydreaming about the practical advantages of having white men and women living with his tribe, such as their cattle and corn. He did not want missionaries from the Bay of Islands, taking up time with their praying and singing. He wanted a few whites with good, old-fashioned fire in their bellies and guns in their hands, who would be stout protectors of any land that Te Puni and Te Wharepouri let them occupy.

The following day, Saturday 21 September, Wakefield launched his campaign. Barrett, presumably under instruction from Wakefield, 'conferred at Pito-one [later Petone] with its associated chiefs, Te Wharepouri, Te Puni, Te Matangi, Mahau and . . . others', asking them to 'have a meeting of all the inhabitants of Port Nicholson, to discuss selling the land . . . for the Europeans to come from England among them'.[26] However, Te Puni and Te Wharepouri suggested pertinently that Wakefield should explore the land he was interested in before discussing transactions. Accordingly, while Rawinia and the girls took Jerningham to see their extended family at Pito-one, Barrett and William Wakefield canoed up the Heretaunga River with Mahau. These were far from being the only two small craft on the waters. All that day and week, the flash of paddles and ripples of canoes on Whanganui-a-Tara headed in many directions like schools of kahawai, as news of the *Tory* passengers' hopes and expectations was carried from pa to pa. Pito-one village became the focus for dozens of chiefs from most of the other kainga.

As Wakefield, Barrett and Mahau returned from exploring the river and delta, Wi Tako, visiting Pito-one from Kumutoto to keep track of developments, asked Tiki what Wakefield had thought of their land. Although Wakefield wrote in his Journal that he had seen the valley as 'extremely fertile when the river shall be contained and capable of cultivation', Barrett, presumably quoting other off-the-cuff comments that Wakefield had made on the trip, or perhaps being deliberately misleading, replied to the effect that Wakefield was disappointed that there were so many steep hills and so little land for the settlers to grow wheat on, but that the harbour was a splendid anchorage. This last remark became a time-bomb, when Wi Tako later stated, presumably on the strength of this, that he had not sold land — only anchorage rights (although he would also swear that he had not sold anything!).

The next day, being Sunday, was to be a time for rest and more familiarisation. Many inquisitive and friendly locals arrived, despite a gale which had risen, to share the ship's church service with the Barretts. When one little flat craft was swamped, some of the *Tory* crewmen launched the ship's boat and rescued the occupants, giving them dry blankets and a place in front of the galley fire, where they made themselves thoroughly at home for the rest of the day. But when, late that afternoon, news came to the ship that Te Rauparaha was threatening to attack Ati Awa at Kapiti, the mien of the *Tory*'s visitors changed. The good humour disappeared and, obviously worried, the men turned into grim warriors and the visitors departed hastily.

The week continued with Barrett constantly on the go, having two languages exercising his tongue and be-devilling his mind. Also, the intricacies of tribal ownership emerged for him to deal with in many discussions; but his grasp of their implications would have been too limited for him to incorporate comments into his translations for Wakefield. Ngaiti was available but the colonel had become disenchanted by this young man. Consequently, '. . . Ngaiti, who spoke perfect Maori and poor English, was supplanted by Barrett, who spoke uneducated English and pidgin Maori.'[27] Even for a trained interpreter, skilled in languages, a day's work is tiring and accuracy can slip. For the unskilled Barrett, confusion and mistakes were inevitable.

His heavy work load continued on Monday the 23rd when he, Wakefield and Jerningham Wakefield set out supposedly to visit all the different settlements round the harbour. Barrett, presumably, was primed by Wakefield with a lengthy rigmarole about the instructions originally given to Wakefield. At each stop Barrett had to assemble all the villagers and give them the same information: Wakefield was bringing many white people to live in New Zealand; he wanted land for them to live on around Port Nicholson; the tribes would live on reserves among the whites; and Wakefield would pay for the land with goods and guns, precious to the vendors. At the same time, Barrett had to '. . . treat them with the most entire frankness, thoroughly explaining to them that . . .' the colonel wished 'to purchase the land for the purpose of establishing a settlement of Englishmen there, similar to . . . the Bay of Islands; or rather on a much larger scale, like the English settlements in New South Wales and Van Diemen's Land' (Tasmania).[28]

Whether they asked or not, he was also supposed to clarify to the villagers that the land would not be 'inalienable, like that which has become the property of the Church Missionary Society', but that 'the Company intends to dispose of its property to individual settlers', and that it will be purchased 'if at all, on the same terms as have formed the conditions of private bargains for land in other parts of the islands'![29] In other words, the land could be put on the open market after they had sold it.

It was a tall order to communicate all that in one day, using a second language, to the eight or ten main villages sited around Whanganui-a-Tara (see p. 111); and his supposed listeners later demonstrated no memory or understanding of anything he may have said. Barrett admitted later that he never tried to explain the intricacies of the reserves. He had said instead that they were to get a certain portion of land.[30]

Wakefield wrote that they visited all the settlements in that one morning; but the pa residents gave differing accounts of who was visited and when during the following week when they gave evidence in 1842 before Commissioner Spain, asked by Her Majesty's Government to investigate the legality of land transfers. Barrett was positive that they had called on Te Aro, and adamant he had told Wakefield that Te Aro would not co-operate. On analysis, the key fact emerges that it was extremely unlikely that Wakefield and Barrett could physically have visited all the villages at Whanganui-a-Tara in one day to explain the New Zealand Company's intentions and seek approval.

What is certain is that on that Monday 23 September, the colonel, his nephew and Barrett, indisputably spent a long period at Te Wharepouri's home, Pukeatua, at Ngarauranga. It was an extremely attractive pa with cultivations, surrounded by virgin bush. They found Te Wharepouri wielding his adze on a 20-metre tree trunk, producing with skill and artistry what was going to be a magnificent canoe. As they admired the chief's expertise, two passing craft called in. They were full of villagers on their way to Pito-one, part of the merry-go-round of the harbour's people, visiting one another, all agog about the fascinating white strangers. Te Wharepouri introduced the slim, reserved English colonel and his nephew — there was no need to introduce Tiki Parete — and explained their land needs. Te Wharepouri spoke to them without translation, of course; and Barrett would have heard enough to check the chief's accuracy; but who could vouch for it on subsequent days?

The visitors entered into debate, of which Barrett seems to have provided full and varied interpretations, apparently well beyond the capacity of a speaker of mere pidgin Maori. The speeches were of a rich content, ranging from accepting to bitterly antagonistic, as translated by Barrett and reported later by Jerningham Wakefield. Among them was Puakawa's. He opposed Wakefield's intentions vehemently for an hour, warning his listeners of European cupidity and pointing out the 'folly' of parting with the new home they had finally acquired 'after the long suffering and dangers of their migration'.[31]

After Puakawa, the dignified, silver-haired Ati Awa chief Te Matangi orated movingly in favour of having white people come. Te Wharepouri took the floor next, striding backwards and forwards energetically in his usual exuberant style, and taking the opportunity to hold forth, mostly about himself and his glory, and claiming that the *Tory* had been sent especially to meet with him.[32]

All this was sifted through Barrett to his attentive *Tory* companions. In one way, his translating task was easier than if he had been in Europe. Tribal custom gave each speaker almost perfect silence, the only interruptions being 'Korero! Korero!', equivalent to 'Hear! Hear!' There were no interjections, as in an English political meeting, and Wakefield mistakenly took this to mean that the listeners concurred. In other ways, it raised new problems of translation. The chiefs' speeches, impassioned, eloquent and lengthy, were of a style totally strange to the Europeans, full of imagery and references to matters and ancestors past, which Barrett was having to précis quickly for his new patrons. Jerningham Wakefield found Barrett lucid and wrote pages about what he heard in that introductory session. William Wakefield wrote soberly but with satisfaction, that '... the close of the arguments ... ended in a decision in favour of the sale ...',[33] a prime example of the colonel's tendency to report to his directors what they wished to hear, and of his ignorance of other tribes' holdings around the harbour.

The close of the argument came when Te Wharepouri's self-congratulations were interrupted by the mouth-watering smell of food. The umu, which the female members in the canoes had been preparing for the sizeable crowd of sixty or so, had been opened, and the hungry listeners left the chief and went over to the meal. Barrett continued his custodian's role, when the colonel and his nephew had eaten enough, making sure that his fellow-countrymen followed tribal etiquette by taking back to the ship both the remains of their meals and the leaf 'dishes' on which they were served. As Jerningham Wakefield commented, 'A compliance with this custom would cause some astonishment at a large London banquet.'[34]

The next day, most of the same people turned up at Pito-one, plus a sprinkling of chiefs from Pipitea, Kumutoto, Raurimu and Kaiwharawhara. They came to be entertained by the spectacle of Te Puni and Te Wharepouri bartering over Pito-one and Ngauranga with the colonel. No one from Te Aro was present. Apart from the fact that Te Aro chiefs did not feel very comfortable among Ati Awa, they had heard that Wakefield wanted their land, and they were totally against him.

Again, there was a vast amount of detail for Barrett to deal with, but because of the different dialects used by the tribes living around the harbour, the accuracy of his translations was suspect. Warrior speakers were intent on acquiring the guns they had heard being offered in exchange for land. They wanted them as protection against further attack from — it must have seemed to the Englishmen — all sides. Some spoke of Ngati Kahungunu, Waikato, Ngati Tuwharetoa and Ngati Raukawa as being threats. Some saw trouble imminent from Taranaki, Ngati Ira (who were simmering down at Turakirae, after having been shunted further and further south as the northern tribes poured into Whanganui-a-Tara) and Ngati Ruanui; and some mentioned challenges from within their own iwi and hapu. Others were looking for the positive aspects. Despite Puakawa's predictions, they saw Wakefield as dignified and chiefly, and thought it would all be good. Barrett even found himself and Love 'held up as an example of what could be expected from the new Pakeha'.[35]

At the end of the korero, the sound of loud agreement greeted Te Puni's question seeking approval for a sale. Wakefield, hardly able to credit his luck, asked the chiefs, through Barrett, if they had indeed made up their minds. They, in return, asked him if he liked the place, because they had decided to accede to his requests '... on their own judgment, aided by the advice of their people in the neighbourhood'.[36] The colonel took this vague response as a positive one and decided confidently that of course he liked the place, that this had been a well-conducted exercise, and that he now all but owned Port Nicholson. His journal entry for the day exudes his self-satisfaction at the successful conclusion of the talks and the way that the chiefs had all shown liking and respect for him. His timetable was starting to look healthier, even as his rationality became suspect.

But Te Puni had asked for the 'ayes', not the 'noes', that day, and Wakefield had again heard what he wanted to hear. His despatches report 'the dissent of some grumblers, who owned but little of [the land], and whose only argument against the sale was, that the white people would drive the natives away, as in Port Jackson.'[37] He ignored the fact that some pa had taken no part in the agreement, although both Dorset and Dieffenbach noticed and later reported different and important aspects of this transaction. Dorset sensed that Barrett and Wakefield had negotiated the sale of only Pito-one that day; while Dieffenbach decided that the transaction 'gave the purchaser permission to make use of a certain district'. He, a German, had become aware of tuku whenua. They, and Barrett, queried the legitimacy of the procedures, but did not, or could not, face a confrontation with Wakefield, who appeared to have the ability to freeze unwanted suggestions from his subordinates.

Instead, unchallenged and thrilled to bits, Wakefield carried the day undisturbed. He spoke to the assembled chiefs, presumably through Barrett, and 'begged them to go on board the ship tomorrow, when I would let them see what I would give for the land'.[38] Although Wakefield was aware that other Europeans' arrangements with the tribes had been only temporary, he was convinced that his arrangements would be permanent and quite different from any previous deals. Barrett again reported 'grumbles' to him, from Moturoa and Wairarapa at Pipitea, along with those from chiefs at other villages. The Principal Agent mentioned this to his directors in despatches, but took no other action, confidently expecting the tribal opposition to fold when they saw the goods he had to offer.

Barrett was a seaman hobnobbing with a colonel, and may not have had the confidence to step out of line and insist that his words be heeded; which was unfortunate, since those opposing the deal remained suspicious of the colonel's requests and Te Puni and Te Wharepouri's motives. Unwittingly they became part of the canker which caused themselves, Wakefield, the New Zealand Company and the future government of the country endless tribulation.

*Chapter 5*

25 SEPTEMBER–4 OCTOBER 1839
CHAMPAGNE, GUN SALUTES AND OPPOSITION

---

With blinkers firmly in place, Wakefield pushed ahead to confirm his land acquisition by negotiating a price and having a deed signed; but this next step, which seemed so simple, took him another frustrating three days. Firstly, on Wednesday 25 September, it took hours for Doddrey to bring up from below decks on the *Tory* the 'purchase price' goods, from which the colonel intended to make a selection; and by the time all the boxes and containers and other paraphernalia were on view, the number of visitors was so great that there was no spare deck space where Wakefield could unpack everything. Barrett also was concerned in case fighting broke out over the merchandise when it was exposed; accordingly, after consulting with Wakefield, he told everyone to go ashore and wait until he sent for them the next day.

When everyone returned late the next morning, after Wakefield had unpacked a tempting variety of wares, an embarrassing situation erupted. Two women, who had stayed the night on board, claimed loudly that the greater part of 'the Trade' was still below. Of course it was. The bales and cases everyone had seen the previous day, from which Wakefield had made up his 'payment pile', contained enough goods for all of the land purchases Wakefield might make, not just in Port Nicholson.

Te Puni and Te Wharepouri reacted immediately with physical threats and loud abuse, accusing the colonel of all sorts of deceits. Wakefield, acknowledged for his courage if not for his sagacity, stood firm, but appealed to Barrett, who advised him to take no notice and just wait. Eventually, the two chiefs did simmer down, '. . . shook hands and said it was only talk'.[1] Te Puni later admitted that when it came to the point, he could not turn his back on all those guns and blankets.

The 'payment' was heaped in a jumble on the deck — 100 red blankets, forty-eight iron pots, 100 tomahawks, one case of pipes, twenty-four spades, fifty steel axes, twelve dozen shirts, twenty jackets, twenty pairs of trousers, sixty red nightcaps, twenty dozen pocket handkerchiefs, twenty-four slates, 200 pencils, ten dozen looking glasses. . . ; with the *pièces de résistance* being, of course, the cases of muskets, twenty-one kegs of gunpowder, fifteen fowling pieces, casks of ball cartridges, and 100 cartouche boxes.

Despite this open temptation, the debate between those for and against passing over land to Wakefield continued, as well as the question of how the goods should be divided. The chiefs quite quickly agreed that Te Wharepouri should be trusted with the latter task, but the other argument continued until sundown. Puakawa was still speaking out in opposition; and another unexpected fly in the ointment was Reihana, a missionary teacher acting as caretaker of the land that Bumby and Hobbs had earmarked for Wesley earlier in the year.

The CMS had actually converted Reihana to Anglicanism while he was a slave at Kororareka, and had christened him Richard Davis, but he had changed his allegiance to Wesleyan when he was freed, so that he could travel back to his home at Whanganui-a-Tara with Bumby and Hobbs. Although ambiguous about his religious persuasion, he was adamantly against the New Zealand Company. Wakefield expected him, as a Christian, to support the transaction and give it legitimacy, but the move backfired. Reihana could speak English fairly fluently. He did not need Barrett's translations, as he protested vehemently that some of the land Wakefield was claiming was absolutely not available.[2] Wakefield ignored his outburst and eased him off the ship. Jerningham wrote later that Reihana was 'so importunate for presents to himself . . . so totally devoid of influence or authority among the chiefs, that we did not regret his returning to tend a sick child at home'.[3] This may have been a definite case when Jerningham (and perhaps his uncle, too) was prevaricating.

By Thursday evening, Wakefield had waited long enough. Through Barrett, he told the crowd that there was to be no more debate and no haggling — the goods on display were all that they could expect. Then, still trying to avoid possible disputes, Dicky announced firmly on the colonel's behalf that the goods could not be taken away, but would be delivered to each settlement. Again, everyone was sent ashore.

The next morning they were back, and Te Wharepouri began sorting the 'payment' into piles. After a while, he asked Tiki for a sixth case of muskets. There were only five cases of guns for the six heaps he was creating, and each kainga would want an unopened box of twenty firearms. The colonel saw this as haggling. 'Mr Barrett, I will not give any more,' he said sourly, and went below. The atmosphere was suddenly electric, and could have worsened when, after a period, he sent up a brusque command for everyone to leave the ship and 'for Mr Doddery (sic) to take an inventory of the things and put them below again';[4] but Barrett chose not to tell them to leave, nor to ask Doddrey to put the things away.

Everyone waited in suspense for another half an hour. Te Wharepouri asked Tiki, 'What is the old man thinking about?' 'Considering,' Barrett ventured, suggesting further that this might actually be the way that Te Wharepouri himself may have acted, if he thought that he was being two-timed.[5] Te Wharepouri was silent again. The colonel put in an appearance on deck and said, 'Mr Barrett, if I give them this case, do you think they won't ask for another?'[6] Dicky did not commit himself. The colonel disappeared again and the tension continued. Finally, Wakefield capitulated and ordered another case of muskets to be brought up. Everyone relaxed in relief, both races gave a hand to hoist it, and Te Wharepouri completed his task. Wakefield reported simply to his directors: 'Friday 27 September. Some little delay' over a sixth case of muskets, which he eventually produced.[7] Young Jerningham wrote that his uncle produced it 'at once'.

Te Wharepouri had created equal piles for Pito-one, Ngauranga, Waiwhetu and Kaiwharawhara, one to be shared between Pipitea and Kumutoto, and a smaller one, to confirm his contempt, for Te Aro. This was set aside to be delivered by boat, even though Barrett had told Wakefield again that Te Aro were not taking part, in fact that Te Aro had ignored the whole business; but, again, his words were ignored. Wi Tako, who happened to have come on board to be part of the goings-on, saw a 'good suit of clothes . . . in the pile' set aside 'for his father, the chief of Pipitea and Kumutoto', and dressed himself on the spot.[8] Barrett hurriedly jumped up and announced that no one could have anything until they had signed the document that Jerningham was preparing.

At about this stage Wi Tako's senior 'uncle', the chief Wairarapa, came on board to return a small boat of the colonel's he had borrowed; but, he told Tiki, he had also come to act for the people of Pipitea, because Moturoa, his brother, would not and neither of them was willing to part with their land. This confirmed Barrett's intimations to Wakefield; but Wairarapa, too, was ignored. Wi Tako had peeled off the clothes amiably, but, because he had previously said that Pipitea would not let land go, and, probably, because of his uncle's words, he thought that the Pipitea message had been recorded and later took back to that pa the articles Te Wharepouri had designated for them. Maybe he thought Wakefield could have use of the land under tuku whenua. Because he had tribal connections with both Kumutoto and Pipitea, he had further confused the import of Wairarapa's statement, and did not realize that, to the English, his acceptance of the goods meant Pipitea's acceptance of Wakefield's wishes.

Jerningham, meanwhile, had been busy drawing up an elaborate deed. The possibility of walking the boundaries, as Wakefield had been instructed by his principals,[9] was mentioned and discarded, perhaps due to pressure of time; but at this stage one of the principal Ati Awa chiefs stood on the deck and intoned names as he followed with his finger the summit of the mountain ranges '. . . from the Rimurapa to the Turakirae, and from the Tararua to the sea'.[10] Some historians suggest this was a statement to impress the colonel of the chief's importance and to nullify the power of Te Rauparaha; but Jerningham thought that it was a statement of ownership, and inscribed these physical features into the deed. Ngaiti was there, as Barrett was doing his usual translations for whoever was listening, and didn't announce any disclaimer.

Barrett had previously made Jerningham aware that the 'Kafia [sic] tribe' claimed this vast area, and that young scribe covered any contingencies by stating that Wakefield was buying 'the rights, claims and interests of the contracting chieftains, whatsoever they might be, to any land, whatever . . .' within certain boundaries. 'Kafia', and anyone else for that matter, was included, willing or not. When completed, the copperplate masterpiece was laid on the capstan of the boat, and the mixed assemblage listened with 'great attention and decorum',[11] as Jerningham slowly read it in a loud voice, with frequent pauses. He had included all the 'Be it therefore knowns', 'aforesaids', and repetitious legalese that he had found in the official documents with which they had travelled:

First Deed of Purchase from the Natives,
dated 27 September 1839
Be it therefore known . . . have this day sold and parted with all our right, title and interest in all the said lands, tenements, woods, bays, harbours, rivers, streams and creeks, as shall be hereafter described . . . as a full and just payment for the same . . . to prevent any dispute or misunderstanding. And to guarantee more strongly unto the said William Wakefield, his executors and administrators in trust for the said governors, directors and shareholders of the Company . . . their heirs, administrators. . . . The whole of the bay, harbour and district of Wanga Nui Atera, commonly called Port Nicholson, situate on the northeastern side of Cook's Strait in New Zealand. The summit of the range of mountains known by the name of Turakirai, from the point where the said range strikes the sea in Cook's Strait, outside the eastern headland of the said bay and harbour of Wanga Nui Atera, or Port Nicholson, along the summit of the said range, at the distance of about twelve English miles, more or less, from the low water mark, on the eastern shore of the said bay or harbour of Wanga Nui Atera, or Port Nicholson, until the foot of the high range of mountains called Tararua, situate about

*Bird's-eye view of Wellington Harbour, by Charles Heaphy. The small island of Matiu is easily seen, where Te Wharepouri lived before he moved to Ngauranga.* ALEXANDER TURNBULL LIBRARY

forty English miles more or less, from the sandy beach at the north eastern extremity of the said bay or harbour of Wanga Nui Atera, or Port Nicholson, is the eastern boundary of the said lands, tenements, woods, bays, harbours, rivers, streams and creeks. From the point where the said eastern boundary strikes the foot of the aforementioned Tararua range of mountains, along the foot of the said Tararua range, until the point where the range of mountains called Rimarap strikes the foot of the said Tararua range, is the northeastern boundary of the said lands, tenements, woods, bays, harbours, rivers, streams and creeks. From the said point where the Rimarap range of mountains strikes the foot of the Tararua range, along the summit of the said Rimarap range of mountains, at a distance of about twelve English miles, more or less, from the low-water mark, on the western shore of the said bay or harbour of Wanga Nui Atera, or Port Nicholson, until the point where the said Rimarap range strikes the sea in Cook's Straits, outside the western headland of the said bay or harbour of Wanga Nui Atera, or Port Nicholson, is the western boundary of the said lands, tenements, woods, bays, harbours, rivers, streams and creeks. From the said point where the Rimarap range of mountains strikes the sea in Cook's Straits, in a direct line to the aforesaid point where the Turikirai range strikes the sea in the said Cook's Straits, is the southern boundary of the said lands, tenements, woods, bays, harbours, rivers, streams and creeks. Be it also known, that the said bay, harbour and district of Wanga Nui Atera, or Port Nicholson, does include the island of Wakaroa, and the island of Matiu, which islands are both situate in the said harbour of Wanga Nui Atera, or Port Nicholson, as well as all other lands, tenements, woods, bays, harbours, rivers, streams and creeks situate within the aforesaid boundaries, and now sold by us, the aforesaid chiefs, to the said William Wakefield.[12]

According to both Wakefields and Barrett, the latter translated during the pauses that Jerningham provided, and Ngaiti 'was a subscribing witness, and occasionally explained the nature of the deed as relates to the reserves of land'; but none of the chiefs present at the time could remember this three years later in Commissioner Spain's Court. This is not surprising. Dicky himself had been completely flummoxed by Jerningham's script. As he said later: '. . . I mean, the language. . . .'[13] In Spain's court, Barrett had to translate the deed again, in the presence of George Clarke jnr, Protector of Aborigines, who was fluent in both Maori and English. The deed contains 1,600 words. Dicky managed only 115, and meaningless words they were:

Listen Natives, all the people of Port Nicholson, this is a paper respecting the purchasing of land of yours, this paper has the names of the places of Port Nicholson, understand this is a good book. Listen the whole of you Natives — to write your names in this Book — and the names of the places — are Tararua continuing on to the other side of Port Nicholson to the name of Parangarahau; it is a book of the names of the channels and the woods, the whole of them to write in this book people and children the land to Wairaureki, when the people arrive from England they will show you your part. The whole of you.[14]

Afterwards, Clarke referred to Barrett as 'a decent fellow enough among men of his class', but rubbished his effort, referring to him as 'ignorant', and speaking 'whaler Maori, a jargon that bears much the same relation to the real language of the Maoris as the pigeon [sic] English of the Chinese does to our mother tongue'.[15] Yet, put on the spot by the convoluted language, Dicky could have thought that he had communicated the kernel of the deed's meaning constantly during the previous week, and that the assembled chiefs knew well enough what Wakefield was proposing. They would not have understood the deed better, even if he had been able to put it into legalese Maori.

At the end of Jerningham's reading, Dicky began hustling the chiefs to come and put their marks on the parchment. They reported later that he had tossed out suggestions such as: 'Come and sign — the Queen will see your name', or 'it's only water that you're selling', or 'if you sign, the gentlemen in England who sent out the trade will know who are the chiefs'. Whether he was looking to get as much support for Wakefield as possible by friendly blandishment or by deliberately misleading them is open to interpretation, but he was confident enough of his intentions to repeat in Spain's court in 1843 what he had said then, in 1839.

Responding to Tiki, and the enticing merchandise on the deck, those inhabitants of Whanganui-a-Tara present, young and old, came up one by one and made their marks. Wakefield wrote that Barrett told him they were attracted not so much by the goods, as by the prospect of having white people living with them; but they may both have been saying and hearing what the other wanted. Rawinia and the children are not mentioned in accounts of this period. Presumably, they were on the ship with the assembled multitude. Tribal custom had everybody taking part.[16]

William Wakefield's journal entry reads in part:

> Thus has terminated in the most satisfactory manner this first and most important purchase for the Company. Some years ago I was present at the execution of a deed at the office of the Company's Solicitors, by which a noble Lord . . . disposed of a valuable estate to another large landed proprietor . . . and, with the exception that the purchase money was transferred in the adjoining bank and that all the talk had taken place amongst the lawyers before the deeds were drawn, this transaction has been not less amicably and loyally carried on and agreeably concluded![17]

Some chiefs were satisfied, some were not, but they made off anyway, with those precious commodities weighing down their canoes — all except for Te Aro, of course. Their smaller pile, made by Te Wharepouri for Te Aro 'taurekareka' was delivered later.

That evening, Te Puni and Te Wharepouri, formally dressed in newly acquired suits, joined the upper deck for a celebratory dinner. For Barrett, who had been working 12- to 16-hour days at Wakefield's beck and call, being husband and father to his wife and family, as well as mentor, relative and friend to Ati Awa constantly around him, there was no relaxing yet. This was just another night's work.

According to Barrett, on Sunday 29 September, he made sure that all and sundry received a message that the colonel wished 'to experience a war dance and a pig killed and cooked the maori way'. According to Wakefield, the chiefs organized the haka and hangi themselves, as their celebration; while Jerningham recorded that his uncle 'went round the harbour . . . to invite them'.[18] To be concerned about who arranged this get-together seems unnecessary; and yet the disparities suggest the striving for self-justification or one-upmanship which were to affect the relationship between the New Zealanders and their visitors. In more conflicting reporting, Wakefield recorded that he and Te Wharepouri went visiting Te Aro, whose 'beautiful piece of flat land' he had his eye on; it was his first choice for the New Zealand Company's first settlement. However, Jerningham writes as though he was there too, but neither refers to Barrett. It seems most likely that Dicky went along, but that, by now both William and Jerningham Wakefield ceased to mention his presence, it being taken for granted. They simply wrote down the day's affairs as direct communication, rather than as received interpretations.

Te Aro village was occupied by hapu of Taranaki iwi and some Ngati Ruanui. Both groups had gradually migrated south, away from Tainui might, like so many other heke, and had assisted Ati Awa at Haowhenua on the way. As a reward, when they moved on to Whanganui-a-Tara, Ngatata-i-te-rangi, Wi Tako's father, had permitted them to settle between the Te Aro and Waitangi streams.[19] As Te Aro's inhabitants came to the beach, curious to see the white colonel, Te Wharepouri directed proceedings. He made Wakefield sit in a canoe facing the people, then he harangued them in a supercilious manner, telling them how lucky they now were, as 'taurekareka' (had Te Wharepouri adopted Barrett's earlier translation, or was Jerningham Wakefield just perpetuating the mistake in his text?), to have muskets.

One of the tribal so-called mihinare living at Te Aro challenged him, not at all cowed by Te Wharepouri's arrogance. He accused the Ati Awa chief of trying to do a deal with land that had been set aside for missionary work, and on which Reihana and the other 'missionary delegates' had already built a chapel and houses.[20] But this was the area that Wakefield specially coveted; and when Te Wharepouri rebuffed the interjector, the colonel, remembering his instructions to keep on the right side of the missionaries, suggested placatingly that he should pay something extra for it. Te Wharepouri reacted indignantly. He had arranged the sale. Te Aro had no right to go against him, and, anyway, Wakefield had already paid for everything in the area.

The main Te Aro chiefs were Hami Parai, Marangai, Te Ngahuru, Mohi Ngaponga.[21] Toko was another, who created some confusion in Spain's court later, since his name was similar to Wi Tako's. They had had no intercourse with the colonel, had no intention of parting with any of their land, and appear to have just listened sullenly and without reply or any rebuttal to Te Wharepouri's posturings. This was not unusual korero behaviour, but Wakefield took it to mean that they basically had no objections — despite what Barrett had told him.

Wakefield protested no more,[22] but Barrett later had to undertake ticklish negotiations to obtain this area, because of this paradoxical interaction of different cultures. The Principal Agent subsequently wrote to his directors with euphoria, assuring them that he had had a triumphant reception at nearly all of the pa. Wakefield's self-congratulations pulled wool over his own eyes as well as his board's, and he adopted an air of surprise and hurt later, when the inevitable repercussions dogged him.

Festivities were held the next day, 30 September, at Pito-one. Some warriors from as far away as Wanganui had come to take part, but the disapproving ones around the harbour, notably those from Te Aro, were conspicuous by their absence. Te Wharepouri made various excuses for this — they were at their distant gardens, they were on an expedition, the weather was too bad. Wakefield appeared not to detect the enmity developing in some of the pa against the white strangers; and Barrett, too, although popular with hapu of Te Ati Awa at Pito-one and Ngauranga, had no long-term admirers among the other villages.

Luckily for Wakefield's peace of mind, local Ati Awa supported the celebrations in a big way. They had cut down a small tree that morning and erected a flagstaff near Pito-one pa, supervised by the *Tory*'s carpenter. They then came to the party carrying nearly every musket from Te Wharepouri's 'payment' piles, and wearing their new red blankets, waistcoats, petticoats, shirts and shot-belts, along with the usual ornaments of feathers and 'handsome mats', and their new umbrellas. Te Wharepouri took every eye, wearing a voluminous hussar cloak lent to him by the colonel and flourishing a handsome greenstone mere.

As Wakefield, Dieffenbach, Dorset, Chaffers and the rest of the official party landed from the *Tory*, the crew hoisted the ensign on the ship's mast and fired a twenty-one gun salute, and the New Zealand flag was broken out on the new flagstaff. Hangi had already been laid down for a gargantuan

# WHANGANUI-A-TARA ~ PORT NICHOLSON ~ 1841

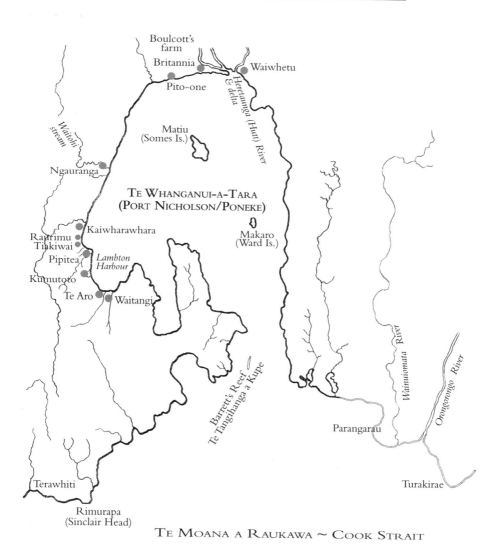

*Te Whanganui-a-Tara is less formally referred to as Whanganui-a-Tara, and Te Moana a Raukwawa as Raukawa Moana, or Raukawa.*
ADAPTED FROM LANDS & SURVEY DEPARTMENT MAP, 1928.

feast, champagne was brought from the ship, and the fun began.

Te Wharepouri and Taringakuri led opposing parties in Wakefield's requested 'war-dance', which Dorset thought was carried out to celebrate selling their land. When they repeated the performance, many of the women joined in with as much gusto as any of the men. Finally, 150 enthusiasts performed an ear-splitting haka, the hangi was opened and the solicited pig, suitably sacrificed, was consumed by all. Healths were drunk, speeches made, and the formal possession of Port Nicholson and district was announced, amid three hearty cheers from all the assembled. Wakefield was on top of the world, the New Zealanders went along with the hoop-la, and Barrett could relax for the first time in days.

During a period of less pressure and worry that followed, Dicky and the others on board made the necessary arrangements before leaving with Wakefield to barter for land in the Sounds. As a result of Barrett's suggestion, the colonel authorized him to send an advertisement across to the *Sydney Herald* in New South Wales. This announced the company's purchases of land in New Zealand, on both sides of 'Cook's Straits', 'with the exception of all previously acquired purchases . . .', to alert people involved who may not be in the country to hear what was happening.[23] Chaffers, helped by Te Puni's eldest son, Te Whare, surveyed and drew up a chart of the harbour; and the colonel briefed an appointed New Zealand Company harbour custodian about where the settlement was to be placed. Barrett, for his part, asked Wi Tako to build him a house, ready for his and Rawinia's return.

The colonel, now champing at the bit to catch up on his schedule, hustled the *Tory*'s extended family into their domicile between decks and, on Friday 4 October 1839, the ship sailed out on its next quest. As they passed harbour features to right and left, Colonel Wakefield, like a king dubbing his knights, proceeded to name them after members and supporters of the New Zealand Company, ignoring their existing tribal names. The last feature so named were the rocks to which '. . . our worthy and honest co-operator'[24] had drawn attention as the *Tory* entered two weeks previously, naming them for posterity Barrett's Reef.

*Chapter 6*

4 OCTOBER–28 NOVEMBER 1839
MIDDLEMAN IN THE SOUNDS WITH
TE ATI AWA AND NGATI TOA

---

When the *Tory* reached Cloudy Bay that evening, the barque *Honduras*, chartered by the Weller brothers of Otakou, was bobbing at anchor in Port Underwood, loading up before sailing to Sydney. Responsible again for station affairs, Barrett arranged for her to pick up his whale oil and bone cargo at Te Awaiti and made sure that Wakefield's despatches were on board too. He had told the chiefs at Whanganui-a-Tara that their names would come before important men in England. His name would be there too, when those despatch envelopes were opened.

Hearing news that Ngai Tahu warriors were at Pireka, on the 1839–40 war-path against Ngati Toa, and that Love was near death, Barrett bundled his family and their companions into a sealing-boat, and set off urgently the next day, promising to come back to Port Underwood in time for the colonel's expedition to Taranaki. Arthur Elmslie and James Heberley (commonly known as 'Worser') were left behind with instructions to guide Wakefield round the Sounds.

Jacky Love died soon after the Barretts' return and Te Awaiti, quiet since the close of the season, came alive with keening and wailing as 200 to 300 visitors poured in and expressed their grief, respect and affection. Ati Awa men and pipe-smoking women, whalers, even children, were all involved in some way with cooking and feeding and generally caring for the visitors to the tangi, at the end of which the European whalers and Te Ati Awa buried Hakirau on the grassy slopes behind the settlement. They marked his grave in a manner usually reserved for great chiefs of the southern tribes, with 'half a canoe, stuck upright in the ground'. Ornamented 'with broad stripes of red and white ochre, and edged all

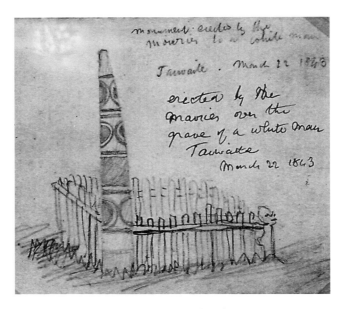

*Jacky Love's memorial was still standing when the surveyor J. W. Barnicoat visited in 1843. He was told it was an important Maori chief's grave.*
HOCKEN LIBRARY

round with a fringe of feathers', it made a striking memorial.[1]

At the same time, many Ati Awa at the station had been quietly preparing for more hostilities, both with a Ngai Tahu taua, supposedly travelling north against Te Rauparaha, and with their enemies of yore, Ngati Raukawa. Te Awaiti Ati Awa were unaware that the chiefs had abandoned the taua precipitately when they heard about William Wakefield's purchase, and were now preparing to sail to Australia to get the authority of their tribal lands confirmed by the Governor;[2] but nothing would prevent the impending clashes at Waikanae, forecast by the people who had come home from Waitohi's tangi at Kapiti to attend Love's at Te Awaiti. They were agitating now for reinforcements to travel back to Kapiti with them to fight their old enemy Ngati Raukawa.

This was the background as Barrett returned to end-of-season maintenance, again arranged a manager for Te Awaiti during his coming second absence and settled Daniel Love back in with his own family. The

*Tory* nosed into their bay on 13 October, its occupants bothered after dealing with warriors from different iwi who had boarded and rampaged round the ship soon after Barrett had left her at Port Underwood. Some warriors had complained vociferously about Barrett being negotiator at Whanganui-a-Tara, while Ngati Toa and Ngai Tahu chiefs were furious, in general, at the way in which Wakefield had omitted to consider them in his negotiations. There had been an awkward period before Wakefield had regained control on board, but now, unfazed, he was impatient to move on to the next stage of his plans.

However, Rawinia was in poor health, so Barrett asked John Brookes, a sawyer at Cloudy Bay, to sail with Wakefield as interpreter. Brookes was reputed to have spent 'eight years among the ... Waikato tribes', and to be 'a good Maori linguist'.[3] Three days later Wakefield set off on his next quest, to barter for land with 'Old Robulla' at Kapiti, accompanied by Brookes. Barrett arranged for two whaleboats to start them on their way by towing the *Tory* to the northern entrance of 'Tory Channel'.

According to Wakefield, when he returned two weeks later, apart from a few hiccups his latest bargaining had gone like a charm. Ngaiti and Brookes had been ineffective (in fact, Ngaiti had left the ship there, 'with all his boxes and goods', to return to his relatives),[4] the negotiations had been 'difficult and disagreeable', and there had been a few altercations, but Te Rauparaha had greeted them as distinguished visitors and Wakefield had eventually achieved everything he could have dreamed of. After discussion with Te Rauparaha and various chiefs, Jerningham had shown them a map of New Zealand, pointing out the areas Wakefield was interested in, and 'the Old Sarpint' had forthrightly agreed that those were the places he was talking about, and had dictated their names. Wakefield had offered Te Rauparaha guns and goods for them, in recognition of his *droit de seigneur*, while telling him sternly and repeatedly that the land would no longer be his, and could never again be sold to anyone. Te Rauparaha and his chiefs had accepted Wakefield's stipulations and his offer of goods, and 'finished by saying, "Look at the land! If it is good, take it!"'. There were all of eleven chiefly marks to prove this,[5] on the document that Jerningham Wakefield had composed for the occasion. Te Rauparaha's understanding of the same meeting was quite different. He had actually been upbraiding Wakefield for dealing with Te Ati Awa, expostulating that Ati Awa had no rights to Whanganui-a-Tara and insisting that Wakefield should have dealt with him. He had not, in fact, been enumerating the names of the areas he was 'selling'; he had been bragging about his conquests.[6] Wakefield was correct in only one aspect: the incompetence of his interpreters — but he did not learn this till later.

While at Kapiti, Wakefield had met and talked with the captain of a ramshackle little brig, the *Siren*, who had on board deeds for agents in New Zealand, sent over urgently from Sydney to confirm earlier pre-paid 'purchases' of land. Wakefield heard that the *Tory*'s doings were causing intense interest in Australia, and that it was rumoured the British Government was about to take action to control land sales in New Zealand. This was an unpleasant goad for the Principal Agent; and, although convinced he had an almost 'safe and binding title' for immense acreage from Te Rauparaha at Kapiti, once back at Te Awaiti, he hustled Barrett to be ready to help negotiate with Ngati Toa and Ati Awa chiefs living in the Sounds region. Wakefield's ambition was to acquire land wholesale by latitude: from 38 degrees, near Mokau and Kawhia, down the west coast of Te Ika a Maui to latitude 43 degrees, at the top third of Te Wai Pounamu, and from 41st to 43rd on the eastern coast.[7] Whether through cunning or ignorance, Barrett agreed to this grandiose absurdity.

At Barrett's whaling station, Ati Awa warrior-whalers were home again after taking part in the expected fighting at Waikanae, and were drumming up additional fighters to take back to the fray. Wakefield sent a message asking them to delay their revenge. He had tempting cargo to show them and an offer to make. Intrigued, they obliged and began trekking over the hills to East Bay, on the north-western side of Arapoua Island, to hear his proposals. The same day, Wakefield with the Barrett family and friends aboard the *Tory*, set out for East Bay — just a short sail round a headland — for the next round of parleying.

As they rounded the point and anchored, they saw the hills surrounding

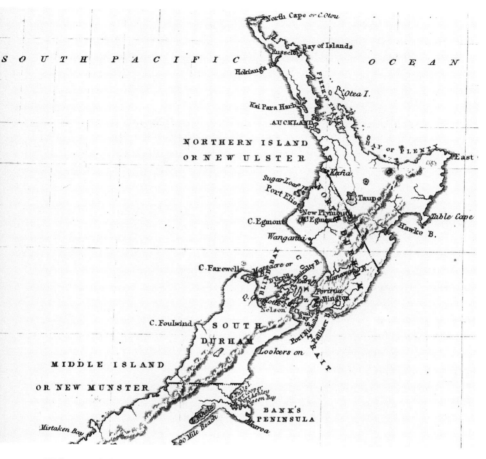

*William Wakefield hoped to acquire land wholesale by latitude, from 38°, near Mokau and Kawhia, down the west coast of Te Ika a Maui to latitude 43°, and from the 41st to 43rd on the eastern coast (shown by dark broken lines).*
DETAIL ADAPTED FROM 13TH REPORT OF DIRECTORS OF THE NEW ZEALAND COMPANY, 1844.
AUCKLAND INSTITUE AND MUSEUM

the expansive bay clothed in a yellow carpet of flowering wild cabbages,[8] and heard the sound of praying and singing from the 300 or so Ati Awa and Ngati Toa already gathered. A welcoming feast awaited all those on board the *Tory*, then after the meal Rawinia and the girls returned to the ship and Barrett and the colonel, for the umpteenth time, began their spiel about land, reserves and total alienation.

Wakefield was under the impression that Te Rauparaha at Kapiti, together with 'these Gnatiawas' (*sic*) and Ngati Toa gathered at East Bay, owned most of the land from the Sounds in Te Wai Pounamu up to Kawhia in Te Ika a Maui, either by hereditary title or by conquest. He thought, therefore, they could do what they wanted with it. In fact, there were over a dozen other iwi with hereditary holdings between Wakefield's latitudes,[9] who would have denied this indignantly. Some had remained in their tribal areas, others had gone into hiding, or vacated their land temporarily and voluntarily because of threats, but none had left permanently.

Ati Awa at East Bay were in this latter category, and were ready to return to their ancestral territory. They were aware of the wealth on the *Tory* and, having heard what their related hapu at Whanganui-a-Tara had received, positively pressed Wakefield to deal with them. They had no concept of alienating the land permanently. They aimed to strengthen the tribe by having Europeans living side-by-side with them at Taranaki, Europeans who would bring bountiful benefits and deter further attacks from Waikato; and they apparently wanted Tiki alongside them there too.

Within two days, Wakefield was noting in his Journal that 'the same reservation of land for Mr Barrett and the children of the late Mr Love, as for the native chiefs'[10] had been stipulated in the discussions. Barrett had had every opportunity, as interpreter, to be the instigator of this clause; and if he had, he perhaps did not realize that the Europeans at Taranaki might be as overwhelming to Te Ati Awa — and himself — as Waikato. Then after six days of atrocious weather, which Wakefield blamed for the delay, the agreement was signed on 8 November 1839.

The deed itself was as verbose as the first, but even less valid and more ludicrous: the constituent places were again written down by Jerningham from chiefs' and Barrett's dictation, but this time Jerningham spelt names in a way that the owners would not have recognized, and mistakenly he actually wrote in one iwi, Ngati Ruanui, as a physical feature of the land being 'bought'. Again, there was no effort made to walk the boundaries.

Wakefield was apparently unconcerned about the responsibilities he was laying on Barrett, or the opportunities he was giving him. He had only two months to complete all of his southern purchases, and he must have been inordinately relieved when thirty chiefs at last made their marks. Over half of the signatories were Rawinia's kin from Puketapu or Ngamotu hapu of Te Ati Awa; and Wakefield neither knew nor cared that none of them was a principal chief.

The subsequent distribution of 'payment' on the deck of the *Tory* sparked a riot such as Barrett had warned Wakefield about at Whanganui-a-Tara. There being no chiefs with sufficient mana to control the crowd, blankets were torn and mirrors and pipes smashed as strong arms and bodies pushed, shoved and grabbed. Most of the whites scrambled into the rigging, and a great deal of damage was done in a few minutes before Barrett, and 'some other white men well known to the natives'[11] stepped into the midst of the heaving throng and gradually restored order — no mean feat considering the 150-strong crowd.

Barrett treated the whole affair as a 'perfect farce', and Heberley commented in his diary, '...There was no writing to show the boundarys or any quantity of land but a certain hill or point so the natives did not know what land they sold, if any other person came and offered any trade they would sell it over again....'[12]

Wakefield had heard about such mix-ups, but dismissed the idea that it could happen to him. Indeed, he wrote glowingly of his achievements to his directors and reported that only two more areas remained to be purchased: 'Waikanai and its neighbourhood; and ...Taitap, including Wanganui, to the southward of Cape Farewell.' He dismissed the shipboard fracas in a few words.[13]

At this point, he also penned an official letter to Barrett, requesting him as 'Agent of the New Zealand Land company in Cook's Strait' to inform all Europeans in the area of proceedings, and to ask the said Europeans to make their services available to the coming settlement. He also requested Barrett and Heberley to stay with him for a while longer. By now, Barrett had been with Wakefield for most of the two months allocated to fulfil his directors' purchasing instructions, was well aware that

Taranaki was in the area that the Principal Agent still had to buy, and that the colonel still had to sail north to Kaipara to inspect, and take over, land that the Company had bought from a previous 'owner'. He chose to remain with his colonel, and Heberley stayed with Barrett.

They headed for Kapiti the very next day, Sunday 10 November, their speed slowed by a contrary gale and the extra weight they were carrying — some Ngati Toa chiefs, to whom they had promised transport home, and a hefty 25-metre trunk of raw timber strapped to the *Tory*'s keel, which Chaffers had felled in West Bay to replace a damaged foreyard.

Once anchored at their destination, they found another a vessel sent to buy Kapiti for Sydney merchants. The *Tory* was just ahead of the eye of the speculative land-buying cyclone, but repairing the foreyard did not speed things up for the now impatient colonel. The baffling wind dropped away completely, and they were becalmed for a further six days before, on 17 November, a light southerly breeze drifted them on as far as Waikanae, where it faded again.

Here, Barrett had to interpret for three Wanganui chiefs who descended on Wakefield, intent on sharing the treasures he had on board. Having described the previous shipboard proceedings as farcical, Barrett cannot have been impressed with the legitimacy of these next dealings. He dutifully gave the usual explanations about reserves and total alienation, as the chiefs described the boundaries of the land they were talking about, signed a deed specially drawn up for them, received 'a fowling-piece each', and 'sold' Wakefield the land from Patea to Wanganui![14]

Wakefield was not unaware. In his journal, he admitted that he only negotiated with these three chiefs '. . . so's not to be entirely idle . . .'[15] while becalmed. He did know it was impossible to complete the bargain until he met with the rest of the tribe at Wanganui; and, to facilitate this, he offered to take Kurukanga, one of the three 'sellers', back there.

Barrett, meanwhile, had gone ashore on unexplained business. (His occasional absences need noting; in some four months, he is supposed to have fathered two children, one from a young Ngai Tahu woman in Te Rauparaha's household.) He had already left the ship the next morning,

*Te Rauparaha, feared for his brilliant skills of improvisation and generalship in warfare and respected for his trading and horticultural skills by all Maori tribes, was denigrated by Europeans as 'Old Robulla' or 'The Old Sarpint'.* HOCKEN LIBRARY

Monday the 18th, when the misunderstandings of Wakefield's initial talks with Te Rauparaha emerged. He heard later how 'Old Robulla' had then arrived again on board, demanded some grog, said that he needed more guns, and offered more land in exchange. Simultaneously, he exerted pressure by intimating he knew that the barque from Sydney, still in the harbour, was also interested in his land. In response to Wakefield's indignant protests, Te Rauparaha had stated firmly that only Taitapu and Whakatu had been included in the previous deal when they last met — the 'Taitap' that Wakefield had mentioned as still to be obtained.

Brookes had been the inefficient interpreter again, in Barrett's absence, and through him Jerningham reported incorrectly that Te Rauparaha was referring to Taitapu and Rangitoto. (No wonder that land titles of the future became tangled!) Wakefield, still through Brookes, had upbraided

Te Rauparaha strongly, accusing him of lying and duplicity and of behaving like a slave, instead of a great chief. Te Rauparaha had listened in imperturbable silence, imperiously called for and drunk another glass of grog, and departed.[16] The white men on board had been dumbfounded. The deed that Te Rauparaha had signed was void. The future appeared in much less of a rosy glow; and, for the first time, it dawned on them that they may not easily obtain the ends they desired. They may have to use some force.

Barrett had returned by that evening. A good breeze sprang up and they sailed north, as Barrett piloted the *Tory* carefully towards Wanganui. Both he and Kurukanga warned Chaffers about the shallow sea up that entire coastline, and Chaffers cautiously anchored 4 or 5 kilometres out from the river's mouth. Barrett delivered Kuru back to his people in the ship's boat, sounding the bar at the river's mouth on the way. He returned to the *Tory* just in time to let the ship scud out safely to sea, under close-reefed topsails, as a sudden, potentially destructive west-north-west squall sprang up. He reported to Wakefield that the bar was only two fathoms at half-tide, making the colonel aware of its unsuitability as a harbour. The squall gave Wakefield no chance to conclude the illustrious purchase from 'Patea to Wanganui'.[17]

Snow-capped Tongariro and Taranaki mountains dominated the breaks in the clouds for the next seven days, as gales from different quarters, gallingly interspersed with calm nights, kept the ship tacking off the coast, unable to approach Ngamotu because of the shoaling waters. Wakefield wrote scathingly about the dangers of the open coastline as he saw it and noted that, despite the reputed qualities of its soil, he could not imagine it being settled 'for some years'.[18] Barrett knew, after the years that he had lived there, that weather conditions were suitable for boat launching and landing for two-thirds of the year, but patience was needed.

During this frustrating period, Barrett entertained all and sundry with tall stories evoked by landmarks he spotted and fantastic experiences connected with Tama-te-uaua. Finally, on 27 November, Chaffers took the chance and anchored the *Tory*, rolling heavily in the big swell, in nine fathoms of water, two miles out from Ngamotu — continuing to be referred to by European sailors as the Sugar Loaves.

Although it continued to blow, Barrett wanted to try for the shore. Leaving Rawinia and the girls aboard, he set off in the ship's whaleboat with the usual crew, plus Heberley, and Te Whare — Te Puni's son — and Tuarau[19] as the 'emissaries' from Port Nicholson, to give weight and authenticity to any initial business talk with the people on shore. It turned out to be still far too rough for them to land, but two figures on the beach who saw them swam out,[20] climbed into the whaleboat and returned with them to the *Tory*. They were close kin of Rawinia and other Ati Awa on board, and a touching scene ensued, as they all met for the first time since Tama-te-uaua had separated them six years before. The Europeans watched in silence as the reunited people sat down on the deck for a prolonged, emotional, wordless period, weeping with their heads buried in their arms. Then they all got up and noisy glee broke out, as the two newcomers greeted their relations ecstatically and hongied with everyone else, captain and crew included.

Afterwards Barrett translated, as one man described in some detail the starvation and casualties arising from their defence of the island of Mikotahi against Waikato — which the Barretts and Loves had heard about at Te Awaiti — and all their deprivations since Tama-te-uaua had left Ngamotu. They had been fearful of all sailing ships, since a foreign vessel[21] carrying soldiers had bombarded the neighbouring iwi further south at Te Namu. They had even been scared of the *Tory*, as they had been of Bumby and Hobbs's missionary ship a few months earlier; and very relieved when they turned out to be men of peace. Their latest European visitor had been on foot — another missionary, Te Wiremu, the Rev. Henry Williams, who had passed this way only a few days previously. He had left mihinare and books and strict warnings against dealing in land with white men.

This was hardly an auspicious start for the colonel, since it sounded as though there would have to be some little persuasion used. At this point[22] discrepancies creep into the reports about what followed. Appearing before

Spain in 1842, Wakefield stated initially that he almost concluded a sale there and then with the two or three chiefs on the *Tory* that first night who wanted whites as protection against Waikato. Later, recalled to the stand that same day, he intimated that many Ati Awa at Ngamotu had begged him to leave Barrett to arrange the sale. Barrett, for his part, gave evidence in Spain's 1844 court that, after he had told the two men who came on board the *Tory* that the colonel had come to buy land, Wakefield had given them a blanket each and they had said that they would sell to him. Then he, Barrett, had taken them back to shore. After he returned to the ship, the wind began to blow up; Wakefield called him to his cabin and asked what his forecast was and when Barrett said 'Unsettled', the colonel instructed him to take his wife and children ashore and complete the buying of the land. Barrett was being cross-examined by William Wakefield at this time, and Wakefield did not query Barrett's answers; but these differences open up the question: Did Barrett engineer his own appointment as sole operator, knowing his employer's time-frame was crucially stretched?

Whatever occurred, Wakefield decided he could make do with a replacement interpreter on his trip north and gave Barrett a month to do the job, Barrett assuring him that by then he would have most of the main chiefs from Patea to Mokau assembled to sign the deed of sale. Rather impatiently, one senses, the colonel watched as Barrett transferred his family into a whaleboat, somewhat overloaded with the addition of Dieffenbach, Heberley, 'Black Lee' and Tuarau, as well as the Barretts' goods and chattels.[23] Thus this ordinary seaman reached the top of the tree as the sole negotiating representative of the New Zealand Company in Taranaki. Wakefield barely waited to see them surf safely through to the beach, before weighing anchor and disappearing north for 'Hokihanga'.

*Chapter 7*

28 NOVEMBER 1839–21 FEBRUARY 1840
BUYING TARANAKI AS SOLE AGENT

After spending an uncomfortable night on Otaikokako beach under the upturned hull of their whaleboat, the tired travellers were 'joyfully greeted' the next day by forty or fifty Ati Awa — the survivors of those who had not travelled south with Tama-te-uaua. Rawinia, Tiki and the mokopuna were embraced with poignant tears and waiata, while Barrett's other companions were given a dignified welcome. Rawinia's father, Te Puke-ki-Mahurangi, was there, but not her mother. Kuramai Te Ra was still held by Waikato.

The Barretts and their companions were given accommodation in the tiny kainga at the foot of Paritutu. Alarmingly, on their second night, lookouts reported fires near Kawhia. The general consternation was contagious, and many stayed awake till morning in case of another Waikato attack, but nothing eventuated immediately from that quarter. Instead, within two days of the Barretts' arrival, a small cutter, the *Aquilla*, anchored off Ngamotu, carrying William White, a defrocked Wesleyan missionary from Hokianga.[1] The colonel's intentions and movements were being swiftly bruited along the country's tracks, and land sharks were on the prowl. White tried to convince the Ngamotu people that Ati Awa slaves at Kaipara had transferred their tribal land rights to him, saying that they wanted Ngamotu to have nothing to do with the people Barrett represented; but the locals gave White short shrift, telling him they would deal only with Tiki Parete. White sailed away, but his visit was disquieting.

The mornings were cool, 44 degrees Fahrenheit, warming up to very hot days of 72 degrees in the shade, but it rained a lot, and nerves remained frayed in the days ahead as slaves freed by Waikato, straggling along the tracks near Ngamotu on their way to homes further south, perpetuated the rumour that Waikato was coming. As a precaution, the *Tory* people began working with the locals, catching and drying fish, digging potatoes and hefting them from the distant forest plots to the shoreline, and killing and curing the pigs running round the village. In case of invasion, everyone took canoe-loads of these provisions and firewood to Moturoa, the farthest out of the Sugar Loaves. Dicky and Rawinia had a whare in a sheltered spot there, on the only level platform of rock about 20 metres in circumference, standing some 30 metres above the sea. All over the rest of the island, right to the summit, little huts were perched in every accessible nook, and they put some of the provisions in these. Some were stored in the many small caves on Moturoa that the villagers had hollowed out over the years wherever the ground was soft enough, blocking the entrances with neat wooden doors.

Barrett took Dieffenbach exploring. They found no sign that the old trading post had ever existed, although the mighty clumps of flax still grew vigorously everywhere, sweeping the ground with their indestructible grey-green spears, and the old cannon from the Otaka siege lay spiked and abandoned on the beach. All of the busy pa, thriving six years before, had fallen into grievous disrepair, only karaka trees and overgrown fosses remaining. The 'rear-guard' now grew their food in small, well-camouflaged distant plots in the denser bush, safe from enemy discovery. The spreading hectares of productive, well-tended gardens had disappeared into a terrain covered in thickets of head-high fern and small trees, raupo swamps and meandering streams, through which they had to make their way when they left the trail. For long hours together, the two men trudged over the 'large sand-hills covered in scrub and roots and fibres'. They reconnoitred the fresh water lagoons behind, still home to multitudes of ducks and eels; and when the German inquired about the strong smell of hydrogen sulphate, Barrett told him with a straight face that a drowned atua was taking a long time to decompose![2]

Dieffenbach eventually covered hundreds of kilometres on his own, being most impressed by the coal and minerals he uncovered in the soil and the admirable qualities of Ati Awa he met. He had become increasingly unimpressed, however, by the choice of Port Nicholson as the company's main site and its land schemes in general and forecast unfortunate results. Barrett, his co-employee in the company, appeared to be unfazed by the German doctor's sometimes outspoken opinions. He continued to do his own job for Wakefield as he saw it, while getting on well with this serious-minded scientist, three years younger than himself, light-years ahead of him in knowledge and education, but still an interesting and stimulating companion.

Barrett was not finding his assignment easy. Since the *Alligator* affair at Te Namu, the villagers' fears of strangers had been exacerbated by the few missionaries who had visited them. Ati Awa described them as fine fellows, full of high talk about a man called Jesus and descriptions of life without war, who had left behind precious papers, testaments and catechisms. But Dicky discovered that the missionaries had also issued stern warnings against all other types of white men, describing them as rewera, who would indubitably cut Ati Awa throats or drive them away.[3] This message was reinforced by the passing ex-slaves. They had all absorbed a sort of Christianity while living with their captors near the missionaries in the north; and the latter had also taught them to be suspicious of all white men who were not associated with the church.

Despite having sent White away scornfully, Ati Awa were affected by this propaganda, and were divided over Tiki Parete's argument that Waikato would hesitate to attack once Europeans were living around Ngamotu. With no charismatic, well-dressed colonel at his elbow, no *Tory* anchored off-shore as a focal point, and no tempting goods on shore, Tiki made no headway at first. Not even Tuarau's support carried weight, and Barrett's father-in-law was one of the most obdurate, actually telling Tiki to go away.

Then as news of Tiki Parete's mission spread, day by day groups of twenty or so curious tribespeople began trailing into Ngamotu — Ati Awa from scattered outposts around Pukerangiora and Taranaki from as far south as their tribal lands to the south. Barrett spoke to them all at a 'great meeting'.[4] No boundaries or payment were settled on and, by the end of the month, only half had agreed to sell. Depressingly for Barrett, when no *Tory* appeared with the promised bounty and deed for them to mark, the hopeful vendors drifted away, leaving directions that they be sent for when the ship arrived.

It is possible that in searching for more chiefs to coerce Barrett walked scores of kilometres. He intimated that he explored Waitara in vain and discovered only a few members of the Taranaki tribe still living around Opunake, whom he tried to interest, but without much response. No traveller brought news that the *Tory* had run into trouble after leaving Hokianga on 19 December and was now lying up the Wairoa River, facing a month or two of repairs; although Dieffenbach wrote to the colonel there, warning that his delay may jeopardize the purchase of Taranaki land if he did not return soon.[5] Nor did they hear that on Christmas Eve, in Sydney, Captain William Hobson, who had just arrived in Australia, had been sworn in by Governor Gipps as the first Lieutenant-Governor of New Zealand.

After Christmas 1839, disturbing news reached Ati Awa at Ngamotu again that a large group was definitely heading south from Waikato country. Gravely, the news was assimilated. About half the band slipped away to hide near cultivations concealed further inland. The other half took refuge on Moturoa with the Barretts, and they all waited in suspense. Then into Ngamotu walked a large group of slaves returning home, and fear turned to jubilation; Kuramai Te Ra was with them for a joyful reunion with Rawinia. Dieffenbach had actually met the group while making his way to Mokau, had spoken with Rawinia's mother, and was able to tell her that her daughter was only a day's travel away.

Accompanying the travellers were Haupokia, a leading Waikato chief, and Edward Meurant, connected with the Wesleyan mission at Kawhia. Both men had good reasons to be there. Haupokia needed to find out what Tiki was up to, having heard about his presence at Ngamotu from White; and Meurant was hoping that his travelling companions would help him acquire land for a Wesleyan station. They appear to have done just that. While Barrett was still struggling to fulfil Wakefield's requirements,

Meurant talked things over with chiefs Poharama and Puke at Ngamotu, quickly and efficiently earmarked 100 acres for the mission, and procured their signatures on a deed, written in Maori and dated 13 January 1840.[6] He departed having given only two blankets and some fish-hooks as utu for the land, with more promised later. Nine days earlier, and still beyond their ken, the barque *Cuba* had anchored at Port Nicholson, carrying the team to survey that settlement for the soon-to-arrive colonists.

Despite the lessening of tension, the Barretts remained on Moturoa. Otaikokakao, the beach opposite the mainland, where the *Adventure*, or *Tohora*, had been lost, was a much warmer playground for swimming and playing than Arapaoa. And the children could help to gather kaimoana or rimurimu from the local reefs, and catch fish and snare birds. These became important for everyone, as the white men's flour supplies dwindled. There was still no sign of the *Tory*, and everyone was increasingly worried about her fate.

Late one afternoon, on 1 February 1840, nine weeks after Wakefield's ship had vanished over the horizon, a sail was sighted, but it turned out to be the *Aquilla* again. On his return from Taranaki, two months before, White had reported on his Ngamotu visit to Haupokia and other Waikato chiefs. They had been outraged. They claimed the area round Pukerangiora and Otaka by right of conquest! Admittedly, they, Waikato, had been talking of a peace-making visit to Taranaki on several occasions since they had been converted to Christianity, but now (White having prodded the embers of their enmity) they would charge down there immediately, and do a bit more persuasive killing to prove their point of ownership.

White moved in. How much better, he suggested, that they sold their land claims to him and avoided casualties. Waikato, after their exposure to missionaries, were in reality keener to talk about war than act on it. Haupokia, therefore, made the initial investigatory visit with Meurant; then, swayed by the prospect of good payment for this troublesome district and by not having to spill any more Tainui blood, they agreed to White's proposal. Thereby, for goods to the value of about £40 as a deposit, and the promise of £1,000 as the purchase price, William White in Kawhia gained rights over most of Taranaki. In his next move, he approached the leader of the mission at Kawhia, John Whiteley, and suggested they should both sail south immediately, since the Wesleyans might have difficulty obtaining possession of Meurant's purchases if the New Zealand Company bought the land. Whiteley could see his point and sailed with him. White thus became the spearhead of the rumoured Waikato invasion — the first to use an unarmed white man as weaponry. It was these two men who were arriving in the cutter *Aquilla* to put as many spokes in Barrett's wheel as they could. Barrett was only saved from acute embarrassment by the incredible coincidence that Jerningham Wakefield and Dr Dorset sailed in at exactly the same time from the north, in a tattered old brig, the *Guide*, which anchored right beside the *Aquilla*.

William Wakefield's non-appearance was explained immediately. The *Tory* had gone aground on a sandbank off Kaipara. After weeks of delay disastrous to his tight schedule, Wakefield had eventually found and chartered the *Guide*. Frantically torn between duties, he had sailed south in her to welcome the first settlers, then sent her back to Kaipara to collect his nephew and Dorset and the purchase goods from the *Tory*'s holds, and take them to Ngamotu.[7] After all this, the two men reported, Wakefield's hoped-for land acquisition up north had turned out to be a complete fiasco.

The *Guide*'s captain was 'fond of grog and sleep', her mate was a doctor 'ignorant of navigation and seamanship', and the crew consisted of the 'worst class of runaway sailors',[8] but the brig carried the all-important goods to exchange for the land. The antagonists in the paper battle for Taranaki land faced one another — White and Whiteley and Barrett, Jerningham and Dorset. None of them had the faintest idea that the first settler ship, the barque *Aurora*, had reached Port Nicholson on 21 January, even before the colonel rushed south to greet them.

The next morning, 2 February 1840, Barrett received a peremptory message from White, informing him that he, White, had a Deed of Sale for the whole district 'bounded by the Whanganui and Mokau rivers', from 'the rightful owners' who, this time, White named not as Ati Awa slaves at Kaipara, but as chiefs of Waikato and Ngati Maniapoto.⁹ The letter warned Barrett that if he persisted in buying this district from the resident inhabitants he and Ati Awa would forthwith be attacked by those former Tainui conquerors, or — and here came a novel claim — Waikato and Maniapoto would seek the protection of the British Government.

Without the support of Dorset (to whom William Wakefield had assigned the role of New Zealand Company agent) and Jerningham Wakefield, Barrett may have felt intimidated, having the might of the British Government thrown at him, but Dorset and Jerningham showed no qualms. Rawinia's relations were also incensed at White's threats and refuted Waikato's claims. Waikato had neither occupied nor cultivated the land for three years, they asserted. That was the only method of achieving tribal rights after conquest. They, Te Ati Awa, were the rightful occupiers. They had encouraged the rest of the tribe to avoid possible obliteration by retreating to the south, but they, the rearguard, had remained. They had kept ahi ka — the tribe's fires burning. Why else had they suffered these years of hardship and fear, than to continue to retain their land?[10]

The day's controversy was overtaken by a strong north-west gale and Barrett sent urgent messages to the captains of both little ships, advising them to up-anchors and away. The *Aquilla*'s captain reacted immediately and was out of danger in no time. The captain of the *Guide* was not so spry, and those on shore were aghast to see the little ramshackle craft in danger of being driven on shore as darkness fell amid thunder, lightning, wind and rain. All the 'payment' goods were still on board, along with Doddrey and 'Worser' (Heberley).

Barrett lit a beacon fire for the ship in one of the protected nooks on the sea side of the island, and then ushered his guests into the only snug shelter he could offer for the night: the empty whata, standing tall on its four long legs in the middle of the rocky shelf. He had evidently had the provisions moved elsewhere. Dieffenbach, Dorset, Jerningham Wakefield and two Ati Awa climbed up the access stairs — a wobbly tree-trunk with notches cut in the sides — for an extremely disturbed night. Apart from worrying about the ship's fate in the storm raging outside, the hut was so small that the five men had to pack in like sardines across its four-foot width. In the morning, to their relief, there was no wreckage on the beach, but also no sign at all of the *Guide* and no hint of her fate.

By now, it had been two and a half months since the *Tory* had deposited the original team at Ngamotu. There was no danger of starvation — no one living in a small tribal village would have to worry about that — and 'Black' Lee, Barrett's long-time companion and cook, made quite a name for himself showing ingenuity in providing tasty meals. But there was every reason for impatience and concern.

At this stage, although at least three of his companions were well-educated men of a superior class, Barrett seems to have slipped into the leadership role. He kept everyone's morale up, and jollied them into good humour with uproariously embellished yarns about his numerous adventures; and, despite his bulk, he clambered up the narrow, precipitous path to the top of the island two or three times a day, in search of that always-elusive sail.

Minor events saved their days from deepest boredom. A fire on the island destroyed one of the huts, as well as possessions precious to the occupants from the village — slates, pencils, catechisms and muskets. Two of the liberated slaves made themselves at home and they all became a captive audience as the new arrivals began preaching their newly imbibed faith with vehemence. And they were onlookers when punishment for adultery was handed out. This began on shore, where the guilty wife was dragged along the beach by the hair and beaten by her cuckolded husband; then relations of both the guilty man and woman stripped the adulterer's house on the mainland of everything movable, before moving on to his hut on Moturoa, '... from which Barrett's attendants brought mats ... and other articles in great glee'. The offending adulterer '... offered no resistance or even remonstrance, according to Maori custom'.[11]

While the five Englishmen were marooned, Jerningham Wakefield prepared a fourth deed, inscribing boundaries with the help of Barrett and three chiefs who came out and conferred with him on Moturoa. Not one had sufficient mana for such a decisive action and the boundaries they dictated stretched right up past Pirininihi, and included land to which they had no tribal rights whatsoever. Nevertheless, this was the area William Wakefield had indicated he wanted, and which Barrett must have arranged to acquire. Ati Awa not included in the korero later said that if they had sold, they had sold only Ngamotu land, from the Sugar Loaves to the Waiongana River, but the cumulative results of this unilateral action were disastrous, leading to serious confrontations over land at Waitara and, eventually, to the Taranaki wars.

The deed was finally ready, but they went on waiting; contact with the outside European world seemed possible only once when, on 10 February, a barque anchored beyond the outermost Sugar Loaf island and sent a small rowing boat to explore. No one on her saw the hurried signalling on Moturoa and, after 'a short excursion', the small boat returned to the ship, which hoisted sail and disappeared northward. It was the *Cuba*, carrying a lawyer, R. D. Hanson, to Kawhia to bargain for more land. He had been sent by Wakefield, who had now returned to Port Nicholson and was realising the shortage of farming space for the hundreds of settlers landing.

Finally, however, the old *Guide* reappeared on 13 February. Doddrey, 'Worser' and the 'indolent' old captain (who had proved himself quite wise, when it came to the test) had weathered the storm by the grace of God, energetic exertions all round, and a slight wind change, but had been driven as far north as Kaipara in the process, and continuing southerly gales had prevented their return till now.

After the long wait, there may have been an element of weariness in Barrett's bones, as he bestirred himself again. All that explaining to do, all those chiefs to collect together again for signatures, and, horrors, another of Jerningham Wakefield's documents to translate. The chiefs from Ati Awa and Taranaki hapu interested in Barrett's offer solved one problem. They appeared from north and south almost unbidden, congregating on the shore and paddling out in small groups to the magnet of the goods in the *Guide*.

Evidence in Spain's court, over four years later, proves how slipshod and headstrong Barrett now became. Jerningham Wakefield wrote afterwards that there was 'some hesitation' over the £500's worth of goods they were to receive, but that the agreement was then just a formality. This was far from the truth. Barrett evidently concealed from Jerningham and the others the hard-line attitude taken again by his father-in-law and twenty-five or so other Te Ati Awa chiefs as the merchandise was displayed. Ati Awa had heard about the double-barrelled guns which had been included in the payment at East Bay, and were not going to co-operate until they received guns too. They still feared a Waikato attack, and did not believe Tiki when he told them there were no muskets on the ship. Barrett tried promising that firearms would be delivered when the settlers came. This was not good enough. They wanted them in their hands now. Barrett said he would sail away with the ship if they did not believe him. Sail away, they said; no guns, no sale. The whole deal was on the point of breaking down when he made the final threat. He would take his children with him if he left. This blackmail about mokopuna finally broke the deadlock and they agreed to go ahead; but they remained highly dissatisfied and, to the day of his death, were still reminding Tiki about the guns.

At least this compromise moved them on to the final stages of the operation. The following day, 15 February 1840, Barrett faced a deed just as convoluted as the previous ones. Jerningham claimed that he read it out, and Barrett said that he stumbled through the translation and the explanations at the same time. His white companions certainly thought he was saying enough to cover every contingency; but that was not what several Te Ati Awa later testified. In 1844, nearly every witness swore on oath to Commissioner Spain that Tiki Parete's translations and explanations had been conspicuous by their absence, and many denied that the deed had even been read out. Barrett later excused himself by saying that during those two months he had covered the ground, anyway, time and again with all those Ati Awa, but did admit that his explanations about Native Reserves were just as vague as they had been at Whanganui-a-Tara.

However, after some blandishments from Tiki, forty-three men and boys and thirty-two women and girls made their all-important marks,

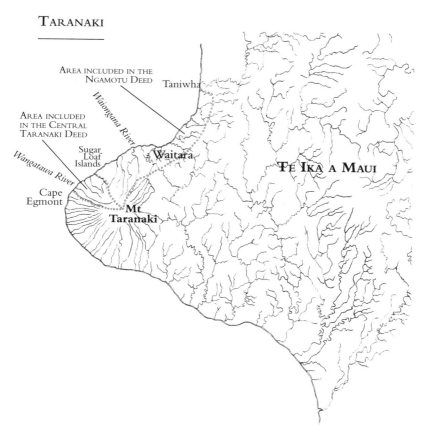

*The two Taranaki areas designated in the Ngamotu Deed and the Taranaki Deed (or the central New Zealand transaction), negotiated for by Barrett and signed for by Dorset on behalf of the New Zealand Company on 24 February 1840.*
ADAPTED FROM THE TARANAKI REPORT: KAUPAPA TUATAHI, WAITANGI TRIBUNAL, 1996.

However, after some blandishments from Tiki, forty-three men and boys and thirty-two women and girls made their all-important marks, Dorset signed as the company's agent, and Dieffenbach, Doddrey and Barrett signed as witnesses.[12] But these Ati Awa were the keepers of the land, 'not the holders of its title'.[13] They had no right to let the land go, and Barrett was later heavily censured by FitzRoy, who took over as Lieutenant-Governor of New Zealand when Hobson died.

Taranaki people, inimical neighbours of Te Ati Awa, had been present but apart during this episode, and now a few who wished to sell their land paddled out to the *Guide* with the New Zealand Company men. Once on board, Jerningham Wakefield drew up another of his memorable documents, using boundaries they mentioned, and they signed it. Barrett made no pretence of having it read and explained to their fellow travellers waiting on shore. They would know what was going on because of all the earlier discussions.

Thus the New Zealand Company became the owner of two areas of land which extended 'from a spot half-way between the mouth of the Mokau River and the Sugar Loaf Islands, to a river called Wangatawa, south of Cape Egmont, and inland, to the summit of the mountain (Taranaki) and thence to a spot on the banks of the Whanganui River, high up its course'.[14] It seemed that only the distribution of goods was required for Barrett to complete his assignment successfully, and the 'payment' for both iwi was landed in whaleboats.

The Ngamotu group divided their goods without dispute, but Taranaki began squabbling and grabbing among themselves as Barrett put their articles on the sand. Almost immediately, a major clash threatened, as Ngamotu saw the in-fighting and ran towards the brawling group, flourishing their arms, intending to help themselves to the Taranaki goods as well if possible. In a flash, the Taranaki men turned with guns, tomahawks and spears in their hands, and the women and children picked up weapons too. They froze into a formidable force, suddenly united to defend their property, their few guns aimed at the advancing Ati Awa. Jerningham Wakefield threw himself down onto a sand hill, to escape the expected crossfire. Barrett, however, reacted instantly, moved swiftly from the Taranaki side across the rapidly decreasing space, 'tripped up one of the assailants and pinioned another'[15] and, together with Heberley, stopped the confrontation and quieted both mobs with some strong talk.

Jerningham rose and dusted himself off, surviving to record it all in *Adventure in New Zealand*; and Taranaki finally departed sullenly with their newly acquired riches. None could have known that on 14 January 1840, the colonial governor in New South Wales, Sir George Gipps —

responsible also for New Zealand affairs — had proclaimed any land sales in New Zealand after that date to be null and void;[16] nor that his subordinate in New Zealand, Lieutenant-Governor William Hobson, had issued the same decree on 30 January, after he had landed in the country. Jerningham Wakefield and Dorset had left the Bay of Islands in the *Guide* just before Hobson's arrival, and the transaction was completed in all innocence. Had the *Tory* not grounded, and the deed been signed as planned one month after Wakefield left Barrett at Ngamotu, the history of New Plymouth may have been very different.

The Barretts, Dorset, Dieffenbach and Heberley took their leave of the steadfast inhabitants at Ngamotu and boarded the *Guide*, along with three visiting chiefs from Mokau. Heberley was to return to his wife and family at Totaranui, and the chiefs wanted to travel to Port Nicholson to offer Wakefield their land too. They sailed from Taranaki on 16 February. Barrett still had several tribes' signatures to obtain, to fulfil his orders from Wakefield to acquire all the land at Wanganui and along that west coast beneath the 38th latitude. However, it was not safe for the *Guide* to call in on the dangerous lee shore, and Wanganui was left for a future date.

Contrary winds once more blew up, and delayed their rounding Cape Terawhiti until the evening of the 20th. With squalls 'whistling off the high land about Sinclair Head' and the long lines of black reefs ahead, the captain was apprehensive, but all on board had confidence in Barrett's navigation. A good moon lit them safely through the heads and daylight found them beating up the harbour to Port Nicholson.

*Chapter 8*

21 FEBRUARY–18 APRIL 1840
THE COLONEL'S RIGHT-HAND MAN

---

The *Guide* anchored in fine weather near Matiu — later to be known as Somes Island — at about 9 a.m. on Friday 21 February 1840. The loneliness of the past centuries was gone. The dark backdrop of the bush was now etched by the masts of five sailing ships, and Europeans were everywhere, looking almost ludicrous — men in trousers and top hats, women in bonnets and crinolines, and children in miniature outfits of their elders.

The *Guide*'s passengers came ashore and explored this instant settlement. Several large prefabricated wooden shelters, full of newly arrived immigrants, and a spacious pataka[1] stood on cleared land which, before the Barrett's departure, had been head-high in scrub. Newcomers had attempted to house themselves, and higgledy-piggledy rows of reed huts and makeshift canvas tents were scattered along Pito-one beach. A jumble of cases, bricks, ploughs, tent-poles, saucepans and other household equipment was piled up among the sand hummocks above the sea, some even below high-water mark. One jetty stood out into the water; but the beach, shallow and exposed to the southerly winds, would not have offered any unloading facilities when the first ships arrived.

Quite a number of Te Puni's people, from Pito-one pa at the western end of the 3-kilometre beach, had come to watch, wrapped in their dogs' hair cloaks and European blankets. Some, stripped for action, were making themselves useful, lifting bales here, carrying objects along the beach there, or helping to construct small reed homes, in return for valued items of European origin. Others sat up on the roofs of houses they had already built, singing as they tied rafters and thatch together with bands of flax. Still other tribespeople had come from further away, and were the centre of attention as they bartered the pigs and kete full of potatoes, which they had transported in the small fleet of canoes hauled up on the beach.

From being one of only 300 whites, nearly all men, living in the whole of New Zealand, Barrett was now surrounded by twice that number of men, women and children in this place alone.[2] The recent arrivals called the place Port Nicholson, as Herd had named it — not Whanganui-a-Tara. The harbour's hapu had shortened this to Poneke. The straggle of shanties and bivouacs was known as Britannia town, and included a 'grog shop', half-way along the previously pristine beach, advertising its position with a New Zealand flag fluttering high. Set up by G. H. Coglan, it had already featured in riotous sprees and become a popular meeting place for sailors, whalers and opportunists attracted to the bay.

There was much for Barrett to catch up on, and many newcomers to meet, most of them looking for excitement and novelty. J. C. Crawford was a young adventurer from New South Wales; Tod, 'a restless character', was another, who had judiciously acquired land from Chief Moturoa at Pipitea during the interim, and had already incurred the Principal Agent's displeasure by building a 'wattle and dab house' on the beach.[3] T. M. Partridge had a tent full of 'goods, furniture, pots and kettles, and all sorts of bedevilment . . .' as a store.[4] John Pierce was selling general provisions (beef, tea, coffee, wine and spirits, ale and porter). George Hunter, who ran a small shop selling general settlers' necessities, had brought out the prefabricated temporary shelters. Carpenters, wheelwrights, watchmakers, and a surprising variety of other occupations, were also offering their services. The number of men carrying guns was explained when Barrett heard that on 10 February, Puakawa, that most determined opponent of land sales to William Wakefield, had been murdered, his head cut off and his heart taken out by the ever-imagined, but so-far-unseen, enemy over the horizon. It had given the settlers a nasty jolt, and they had agitated for, and been handed out arms by Wakefield for their peace of mind.

And Barrett heard more details about the colonel's misadventures in the *Tory*. The whole trip north had turned out to be a huge waste of time and effort. The large tracts of land, to which Wakefield had expected to hold title by now, and for which the company had paid £20,000 to its predecessor, the 1825 company, had shrunk to about one square mile at Herd's Point at Hokianga and '... a couple of small islands in the river called Thames ...'. The colonel had met Bumby and Hobbs, and had taken the opportunity to inform them that he now owned the lands that they had 'tapued' (see Chapter 3); but he had assured them that he would 'reserve a sufficient place for the location of a chapel and mission house ...'[5]

Tiki discovered his friend Te Wharepouri to be downcast and strangely mournful. No one was yet quite sure who had killed Puakawa. It may have been people at Kapiti, Ngati Kahungunu, or any other of Te Wharepouri's numerous sometime enemies, and yet Te Wharepouri was not fired up by the thought of another good fight. Instead, he appeared to be quite overwhelmed by the number of Barrett's countrymen who had poured off the ships:

> I know that we sold you the land, and that no more white people have come ... than you told me. But I thought you were telling lies, and that you had not so many followers. I thought you would have nine or ten, or perhaps as many as there are at Te Awaiti. I thought that I should get one placed at each pa, as a white man to barter with the people and keep us well supplied with arms and clothing; and that I would be able to keep the white man under my hand and regulate their trade myself. But now I see that each ship holds two hundred, and I believe, now, that you have more coming. They are all well armed; and they are strong of heart ...[6]

His heart was 'dark'. His war canoes were ready to launch on the beach, he was going to take his people to Taranaki, and he came to say farewell. Ultimately, it was neither Barrett nor the colonel who dissuaded him — although the latter might well have decided it would actually make the settlers' lives much easier if the villagers decamped wholesale; it was Te Wharepouri's fellow chiefs. The bonanza of goods coming their way from the Europeans was too good to leave. His point made, Te Wharepouri allowed his mind to be changed about leaving. Instead, he appointed himself guardian on the road to Porirua, from whence the enemy might appear, and challenged everyone who travelled it.

The surveying of the embryo town, to which Dicky introduced Rawinia, Caroline, Mary Ann and Sarah, was weeks behind schedule. Captain Mein Smith, kindly, scholarly, religious and diffident, was the New Zealand Company's Surveyor-General. He had arrived in the *Cuba* on 4 January, aware that the first settlers were only a week or two behind him; and that they would eventually number 1,000, each with a land-order entitling the holder to receive one town acre and 100 country acres. Mein Smith had brought sketch plans with him, drawn up in London, showing 1,000 parallelogram sections (as well as a cemetery, market place, botanical gardens, parks, wharves, boulevards, etc.), which were to form the commercial heart of the new town. Smith's instructions from the company directors were to create the whole complex as a unit, with the town sections just one green belt away from the 1,000 country sections, on which the settlers were to grow their food and exports.

The directors had changed their mind since their Principal Agent had left England. Wakefield had been told that the country holdings could be elsewhere, and he had designated the narrow crescent of beach and swampy flat land between Te Aro and Pipitea, with its superb land-locked anchorage, as the town site, and even named it Thorndon.[7] Off his own bat, and since there was no sign of the colonel, Mein Smith decided this was quite unfeasible, and that the Hutt River delta was the only possible area on which he could lay out the settlement according to his orders, and he started his men to work. Wakefield, when he returned, was most displeased that his wishes had been ignored.

Given the confusion over the site, the immigrants in the first four ships had arrived close on Mein Smith's heels to find no sections available for them and no likelihood of any for weeks ahead. The survey at the Hutt was proving to be ticklish and extremely time-consuming. Although there were thousands of hectares of level ground on the eastern side of the river, the immediately accessible ones were around the western side of the swampy delta. The surveyors had to cut through heavy scrub and

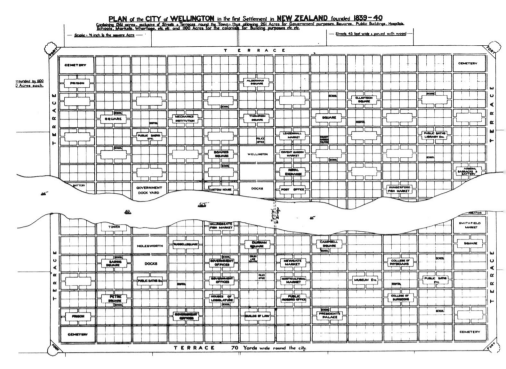

*Plan of the New Zealand Company's first settlement in New Zealand, for which people in the UK bought land-orders. Barrett would have known there would be no room at Whanganui-a-Tara for the promised 1,000 100-acre country sections.*
WELLINGTON HARBOUR BOARD/MARITIME MUSEUM COLLECTION

large trees before they could lay down survey lines, and had to drain swampy areas before they had solid ground into which they could hammer the survey pegs. As a temporary measure, therefore, the surveyors had cleared some land for the settlers 400 metres back from the beach and along the banks of the river, where more of them were 'squatting'. They named it 'Anglionby'. Most of the settlers had been fairly philosophical about the delay up till now, and spent their first two beautifully fine weeks in New Zealand camping out. Wakefield had accepted what Mein Smith told him the directors had said.

The Barretts had no housing problems, however. While Barrett had been buying Taranaki land, Wi Tako, as requested, had built a modest reed whare for them, similar to those Ati Awa were building at Pito-one for the new immigrants. It was on Pipitea pa land, several kilometres south of Britannia (now the eastern corner of Molesworth Street and Lambton

Quay). Dicky put a fence around it and they settled in. Food supplies and whanau support were available, if needed, from their hapu connections at Pipitea and Kumutoto, and means of support for the family would be derived from one or two whaleboats Barrett had had brought over from Te Awaiti to provide a carrying and ferry service. These lay on the beach in front of his house. For Barrett's family, however, this was an utterly alien environment. For him, while much was familiar, it was as though he had spent a dozen years as a Rip Van Winkle.

They were therefore sheltered and dry when, the week after their return, the fine weather broke. Northerly winds and rain lashed the settlement. Cooking in the open became almost impossible, and then the Hutt River overflowed, wiping out part of the surveying work and bringing lapping water and muddy sludge into some of the temporary living quarters. Dealing also with this, as well as long skirts, crinolines and children, tested the women cruelly. Wakefield (for whom Te Puni meanwhile had built a solid little house on the beach) lent his help by moving from place to place among the bedraggled mob, smoking his cheroots, and 'encouraging all to go on and prosper'.[8] Barrett tried to take him up the river in one of his whaleboats, to inspect the damage and boost morale, but could make no headway against the swollen waters and had to give up. Wakefield went up by foot the next day.

There was much unhappiness, discomfort and mutterings. More than a few were becoming unhappy with the colonel's management style, and were already critical that his first choice of Thorndon had been discarded, although a recent newcomer wrote home: 'Colonel Wakefield is decidedly one of the most kind-hearted men in the world and gives universal satisfaction . . . no man can be better qualified for so great an undertaking as the Company's principal agent for New Zealand.'[9] Barrett was not among his critics. He had heard he was expected to be voluntary harbour-master, then Wakefield gave him a boost by writing to his directors:

> Of Mr Barrett's services to the Company I cannot speak in too high terms. I have appointed him Agent for the Natives, which

office will make him the medium between the settlers and their dark neighbours in all disputes and in the allotment of the native reserves in lieu of the land now occupied and cultivated by them.'[10]

A salary of £100 a year came with this part-time position; little enough, when compared with what some others were making, but it represented a steady income to cover Barrett's annual expenses.[11]

\* \* \*

After coping with their first civil emergency of the flood and several minor crimes, the Principal Agent and the people of Britannia decided it was time to organize themselves. On 2 March 1840, therefore, everyone went along to Pito-one to vote for a Council of Colonists. The settlers had been expecting this move, since each of them, just before leaving England, had signed an agreement to follow any laws the company officials might make. They were unaware that government ministers in Britain had declared this agreement illegal, soon after they had sailed; and that those same ministers, knowing the hardships the settlers might experience if they had no means of dealing with illegal behaviour, had hastily despatched Hobson to be Lieutenant-Governor in New Zealand, under Governor Gipps in Sydney. William Wakefield had, of course, also sailed before the embargo was in place; but he typically kept the news to himself when he was later advised in despatches.[12] Therefore, the rank and file had not the remotest idea that their elections would be unlawful. Barrett was not to know it, but William Wakefield and his brother Gibbon made rather a practice of flouting authority. The colonists had no idea even that Hobson had arrived in the Bay of Islands, at the end of January. He had not had time to announce himself and to introduce regulations in the south; and, only the day before, had suffered a stroke. This would, of course, set his intentions back many more months.

As ordered by his directors, Wakefield took over as the first President of the Council and saw to it that Dr George Evans was appointed the magistrate and Mr Samuel Revans the secretary. It seems that the 'old boy' network was busy, since other members elected did not arrive in the country until several days after the proceedings. The council, which his directors had told Wakefield would '... be temporary, until Captain Hobson establishes Colonial Government in New Zealand ...',[13] finally consisted mostly of Wakefield's associates, although Heberley, who had recently moved across from Te Awaiti with his family, and Doddrey were both given official positions, and Barrett was automatically included, as Agent for the Natives.

The names of those elected appeared on 18 April in the first edition of the settlement's first newspaper, *The New Zealand Gazette*. The whole process was puffed up as a 'Ratification' of the agreement in the name of the Sovereign Chiefs, who were granted 'perfect equality', except for the vote.[14] It has been mooted that, since Barrett probably interpreted this article to the said chiefs, they would have understood its finer points as little as those in other documents that the company had produced.

The council summarily issued edicts (already prepared in England by the New Zealand Company directors) and the citizens all accepted that they would submit to the magistrate if disputes arose, would be liable to punishment if they broke the law, etc., etc.; and that the men would be part of the voluntary militia. Everyone worried about their vulnerability after Puakawa's apparently ritual killing, and Barrett, along with every other man in the new settlement between the ages of sixteen and sixty, agreed to be mustered and drilled on a regular basis. The council leaders, intent on remaining on good terms with their neighbours, invited the men of the tribes to train with the militia and proposed that 'chiefs' would be trained to become officers. The latter 'expressed unfeigned satisfaction'.[15]

On 3 March, the day after Wakefield had formed his local government, the lawyer Hanson came back in the *Cuba* from Kawhia. He had not only found Waikato disinterested, he had heard about a new Governor's proclamation that no more New Zealand land should be alienated after 30 January 1840; and that the said Governor — a British naval captain named Hobson — had signed some treaty with the chiefs in the Bay of

Islands. The law-abiding Hanson ceased his business at once, without leaving any sort of land claim, much to Wakefield's displeasure. In this way, the new Council of Colonists learnt what Wakefield had not broadcast to the people of Britannia and backed Wakefield's decision to keep quiet, in case the news should disturb the settlers.

Within these constraints, Britannia expanded. Through February and March houses sprouted at Pito-one almost as quickly and randomly as toadstools. The piles of jumble on the beach, which had greeted the Barretts when they returned from Ngamotu, had been gradually removed into their owners' homes, and been replaced by more orderly stacks of the new arrivals' possessions. A few rudimentary facilities now enabled luggage to be disembarked in a methodical manner, and there was a high demand for private craft such as Barrett's to assist in this.

On 7 March the *Adelaide*, the *Glenbervie* (carrying the company's extra supplies) and the repaired *Tory* met in 'Cook's Strait' and made a rather grand spectacle as they swept up the harbour together at dusk, under full sail and in a sudden tempest of thunder and lightning. Two men among the 180 passengers on the *Adelaide* were to influence Barrett's life to a marked degree — Sam Revans and Dr George Evans.

Revans was a casually dressed bachelor of about Barrett's age, basically a kindly man, but 'somewhat hot tempered, prone to over-optimism in business',[16] a shrewd observer of his fellow men and with a sharp wit. By 8 April, he had brought out the first edition of *The New Zealand Gazette*, and from its four, six or eight A3 pages (depending on the number of advertisements and articles the citizens engendered), divided into several narrow columns filled with heavy black print, came a detailed word image of life at Port Nicholson, colourfully presented and, at times, sharply etched by the leading articles and acid tongue of Revans. Barrett often used its space to advertise.

Dr Evans, a doctor of law, and an assiduous supporter of the previous New Zealand companies, was arrogant, bad-tempered and outspoken, and became exceedingly unpopular. He was to be chief magistrate in the colony although, according to Revans, he had a total disregard for justice;

*Te Aro Pa, whose inhabitants remained in their homes, after the agreement with Shortland, while Thorndon developed around them, but a wharf was built in the middle of the pa's foreshore in 1842.* SKETCH BY E. NORMAN, 1842(?)/ALEXANDER TURNBULL LIBRARY

his wife was almost more unpopular, being '... this old devil ... the cause of all mischief and bad feeling down here'.[17]

Within three days of anchoring, the new arrivals quickly absorbed the disenchantment of the first-comers over conditions at Pito-one and, led by irascible Evans, had turned their backs on Britannia and ensconced themselves at Thorndon. This split the town, and threatened an impossibly expensive duplication of community services. The two areas were kept in touch mainly by a daily mail service. Wakefield called another meeting on 29 March, at which half the land-order holders voted to remain at Britannia, the other half to move to Te Aro. Wakefield made no casting vote, but did pack up his own living quarters and move down-harbour to a position behind the Barretts.

Dicky and Rawinia became aware of Wakefield's intentions when he instructed Barrett to begin efforts to gain clear title to Thorndon land for the company. Apparently, the colonel had no qualms about re-siting the town. As he mentioned ad nauseam in his despatches to his directors, all

*Pipitea Pa, where Wi Tako lived. Its inhabitants gradually shifted out as the settlers arrived. Soon Barrett's Hotel was to be the only reminder of where they had lived – that and Pipitea Street.* SKETCH BY GEORGE RICHARD HILLARD, 1841/ALEXANDER TURNBULL LIBRARY

the chiefs liked and admired him, and he believed that blankets would achieve any of his aims. Barrett, however, was uneasy. He knew well that the chiefs of the Taranaki hapu living at Te Aro had had no dealings on the *Tory*, and he was not sure about the attitude of Ati Awa chiefs Wairarapa, Moturoa and Wi Tako at Pipitea and Kumutoto, and the many chiefs of the other pa. Also, there was another complicated area of extremely unclear ownership — the mission land, over which Reihana (Davis) had greatly confused the issue by changing his allegiance from CMS to Wesleyan and back again. In the process, he made it extremely unclear whether the land he was looking after belonged to the CMS, the Wesleyan Church, or himself; and whether it amounted to 3 acres or 40 or 60.[18] Despite these uncertainties, Barrett was successful in one aspect. By whatever means he used to persuade the chiefs concerned to give Wakefield possession of the chapel built by mihinare — charm, hard talking or 'blanket diplomacy' — he was able to write to Wakefield:

Pipitea March 31 1840

Dear Sir

I take this opportunity of writing to you to inform you that I have this morning purchased the [indecipherable] House at Pipitea I have also talked with the Natives at the Par Taranaki I expect to sail for Queen Charlottes Sound either this Evening or the first thing tomorrow Morning. Please to give the Native as Money one Fowling Piece One Plaid Cloak one Shirt and Trowsers and one Blanket, this is the payment for the Missionary who came from the Bay of Islands as the Land was Tabooed for him and I could not satisfy except by giving them those things.

To Colonel Wakefield, I remain

Your Obedient Servent

R Barret[19]

He told Wakefield at the time that the people of Te Aro pa were ambivalent about whether the supposed 60 acres set aside for missionaries went with this purchase; but that both they and those at Pipitea refused point blank to give up the land, probably about 3 acres (just over 1 hectare), promised to Bumby and Hobbs. Despite this news, either because of optimism over the acquisition of the mission house, or dismay after another overflow of the Hutt River on 30 March, Wakefield a week later took the step of cancelling the surveying at Britannia and starting again at Thorndon.

It is at about this time of March through to mid-April 1840, that Barrett's movements have to be considered. In December, 1840, two boys were born, Henare and Edward, both of whom he is supposed to have fathered from different mothers. There is little of substance known about Edward's beginnings, and links with Barrett are inferred only through some name similarities; but the story of Henare's birth, dismissed as gossip by the descendants of Barrett's surviving daughters Caroline and Sarah, is believed as gospel by Henare's descendants.

Henare's mother was said to be Kararaina, of Ngai Tahu, captured by Te Rauparaha, along with many other young people, on one of his southern raids on Kaiapoi. Although very young, she was selected to be his eighth (or so) wife. Here the accounts begin to differ. In one of them, Barrett was supposed to have eloped with her, and she to have became pregnant with his child. Tiki was said to be absent in Australia at the time of Henare's birth, at which time Kararaina died, and, when Tiki returned, he could not find her, nor could he find out what had happened to her. It is not clear whether he knew she was pregnant. As the story goes, Kararaina's Ngai Tahu relations concealed the baby boy not only from Barrett, but also from Te Rauparaha, who, they say, might well have killed it.[20] When the child was old enough, they carried him secretly to his mother's hapu in Te Wai Pounamu, where he spent his growing years, eventually marrying Louisa Hunter (born 1849) and beginning the Barrett family of the present.

In the other account, Kararaina, pregnant to Tiki, or just having given birth, was released by Te Rauparaha, in the way that so many slaves were released when their captors embraced Christianity.[21] She herself took her baby south and lived on the Taieri Maori Reserve. The baby was called Hone Henare Tumehou (or varieties of that spelling) Barrett.

The question of whether the two men were Barrett's sons awaits further proof; but, given Dicky's responsibilities to his family and Wakefield, as well as his position as Agent for the Natives, it seems unlikely he would have had the time to stray far either from Port Nicholson or Rawinia, and there are no records of his sailing to Australia.

Shortly after the move to Thorndon, a seemingly insignificant event occurred that brought about the Council of Colonists' demise and contributed substantially to the steady erosion of stability for the New Zealand Company and the tribes. After his arrival at Port Nicholson, a Captain Pearson was involved in an unfortunate contretemps over payment with businessmen John and George Wade, who had chartered his ship, the *Integrity*, to transport their stock from Port Jackson, Sydney, to New Zealand. As a result, on 14 April the council, exercising its newly granted powers, peremptorily arrested the captain. Pearson was outraged, escaped back to his ship the next day and within the month had sailed to the Bay of Islands to complain bitterly to Hobson. The Lieutenant-Governor, although still semi-paralysed from his stroke, had been furious at the usurping of his powers. Issuing a proclamation against the council's unilateral action, he ordered Lieutenant Willoughby Shortland to sail to Port Nicholson with a troop of soldiers and policemen to put down the 'rebellion' and 'high treason' of the settlers.²² At the same time he proclaimed the Queen's sovereignty over the whole of the country, effectively pre-empting his emissaries, such as Henry Williams, who were at the time travelling about New Zealand to obtain chiefs' signatures on copies of the Treaty of Waitangi.

However, more bad news reached Port Nicholson, that Wakefield's actions were under scrutiny and that the Governor in New South Wales was forming a land-commission to investigate all purchases in New Zealand. Titles to the settlers' land would be severely delayed, and Barrett and his actions would come under a microscope.

Pearson's escape, meanwhile, became just a nine-day wonder to the weather-beaten, troubled householders in Britannia. The sections that the settlers had expected to find waiting for them when they landed were still not available and did not look like being ready for weeks and weeks. As winter approached, the newcomers continued to live in temporary shelters on the beach and at Anglionby; and were depressed with the daily struggle that got them nowhere, except to keep them alive to go on struggling.

Wakefield, who must have been as despondent as anyone else, was in an unenviable position. He had to accept responsibility for the surveying shambles. When he was an unemployed colonel with a criminal record, the position of Agent for the New Zealand Company had been timely, offered satisfaction and achievement, and given him a finite goal of the acquisition of land for settlement. The follow-up day-to-day duties of being Principal Agent, and the unpopularity he evoked, had not been apparent. Yet, despite being accused of all sorts of personality faults, he maintained a dignified front and wrote long, optimistic despatches to his directors, giving glowing accounts of the industry of the surveyors and the cheerful fortitude of the new settlers. Amongst the latter, he was mainly popular, but several of his company officers were beginning to

*After the Wellington Deed was signed, three large canoes, commanded by Te Puni, Te Wharepouri and Tuarau, took part in a race around the gaily-decorated fleet of immigrant ships. Te Puni took Wakefield in his canoe and came first by paddling under the anchor warps.* ALEXANDER TURNBULL LIBRARY

question his judgement, and to bridle under the coldness of his reserve.

Barrett unquestioningly did what the colonel asked, and the picture is beginning to emerge of him and Wakefield as two of a kind: straight-down-the-line men, obedient to their superiors to the nth degree, and obeying orders without too much moral dilemma. Throughout these days, Barrett appeared content in his relationship with Wakefield and with the warm adulation he received in the pages of the *New Zealand Journal*, the official paper of the New Zealand Company; but he suffered in the long run. As a result of the suspicious tango of moves between Her Majesty's Government and his employer, his family's tribal rights began to be inexorably eroded, in the same way as those of any of the tribes, as they became secondary to European settlers' increasingly voracious and clamorous appetite for more and more land. In the meantime, Barrett made plans to earn a living by fulfilling the day-to-day transport requirements of the people of Port Nicholson.

*Chapter 9*

18 April–12 December 1840
A POPULAR HOTELIER

What the colonists at Port Nicholson wanted most was communication and transport between the two areas they were now occupying, separated by daunting kilometres of impassable land or sea. On 18 April 1840, whaling gear, 'sundries' and eleven of Barrett's whaleboats arrived in town, brought over from Te Awaiti on the decks of the *Cuba*. He mounted the boats on skids in front of his house, ready to carry any light gear or a few people, or to attack whales in the harbour for his own personal profit. The following day the *Harriett* lumbered in from Te Awaiti, carrying 'staves, poultry and sundries; Mr J Wright [one of Barrett's original whalers] and family, and 21 other passengers', the vessel itself being consigned to R. Barrett.[1] He had made a quick purchase of the *Harriett* from Joseph Toms, intending to carry heavier freight between Pito-one and Thorndon.

As at Te Awaiti, Barrett kept an open house with free meals for both races. Jerningham Wakefield felt this hospitality was 'too undistinguishing', in the welcome it gave to 'hungry natives, and idle white men', but gave Barrett credit for being 'Kind-hearted to a fault, always good-humoured and sanguine, and scrupulously honest in all his transactions'.[2] Another settler, T. M. Partridge, wrote about him at that time:

> Dicky Barrett . . . looks as if he had approached the shape of a small calf whale . . . [He] is a great man among the natives, who adore him, and is respected even by drunken whalers. He has befriended many a white man in his districts, and has got the largest heart of any man I know in New Zealand. . . .[3]

The survey team left the Hutt swamps in early April, and started work on the fifty odd acres (20 hectares) at Thorndon Flat. High, rugged hills, thick with forest and scrub and deeply broken by steep ravines, stood at the rear, constant reminders of even more labour ahead. Mein Smith moved in next door to Dicky and Rawinia. He was followed by one household after another from Britannia, and a heterogeneous set of neighbours built up. The Principal Agent of the Company occupied 'a fine residence' on the hill behind, along from where Charles Heaphy and George Moore were building a substantial reed and clay house.[4] The Barretts soon found themselves living cheek-by-jowl on the waterfront with the villagers of Pipitea pa, Evans, the Chief Magistrate, and twenty or thirty other soon-to-be well-known European citizens who clustered to form the commercial heart of the town.

John and George Wade, who had precipitated the arrest of Captain Pearson over the costs of bringing their horses, cattle and goods from Sydney in the *Integrity*, presented as quite well-off and enterprising businessmen from Australia. David Scott, who had lived in New Zealand in the 1820s, had seen Barrett's advertisement in the *Sydney Herald* and returned early in May, to take advantage of any opportunities offering, and to occupy land he claimed to have bought from Pomare in about 1831. He was a Border Scot, canny and talkative, even in Maori, and, since he spoke this with a burr, it 'took some getting used to'. George Moore, slightly older than Barrett, was competing with Dicky in the cargo-carrying business; but in a larger vessel, being captain of the *Jewess*, a 130-tonne 'brigentine' (sic).[5]

Barrett was from the same background as many of the assisted emigrants, but had far more status by virtue of his experience and association with Wakefield; yet the English class-system kept him in his place, well below the 'gentle-folk'. The only 'gentleman' with whom Barrett apparently maintained a close relationship was Wakefield. They were both somewhat divorced from their normal social group. The incoming English people would have found it impossible to relate easily to Rawinia, high-born and proud Ati Awa though she was, despite Wakefield's comments about the acceptability of the Maori wives he had met at Te Awaiti; and Rawinia herself would have felt uncomfortable in their 'social' company.

Perceptibly, as the number of settlers around the Barrett's home increased, so did the atmosphere with the resident villagers become increasingly uncomfortable. Mein Smith began making a considerable impact at Thorndon, with his theodolite and surveying assistants. They were accompanied by a sizeable bunch of helpers, who carried a Gunter's chain, mauls, axes, iron arrows and other cumbersome equipment. The New Zealanders at Pipitea and Te Aro were mystified and entertained at first when these white men started throwing long lengths of linked metal pieces around the unoccupied spaces, sticking iron arrows in the earth, leaning over and squinting through strange boxes on legs, and felling small trees and chopping them into short lengths.

But their amusement rapidly turned to hostility as the survey team then trampled unconcernedly over and around their marae and gardens. Threatening gestures were made and weapons flourished as the surveyors sharpened the short tree-trunk lengths and hammered them into the ground, sometimes in the most sacred and tapu parts of tribal land. And each night, the tribespeople went out and removed the survey pegs. If Barrett was still reminding Wakefield that the land had not been bought, the colonel ignored him, and Barrett would have had no authority over the surveyors. Wakefield now commanded him to distribute twenty blankets, hoping they would be accepted as compensation for the use of the land, but the tribespeople saw them only as bribes to make them stop pulling up the pegs.[6] They continued to express their increasing antagonism violently but without bloodshed as the surveyors continued to abuse their home grounds. Wakefield issued guns to the surveyors, for their peace of mind; but in his despatches gave his directors the impression that the peg removal occurred because the surveyors were working on land 'tabooed' for Henry Williams. He assured the directors he now owned the land, on behalf of the New Zealand Company; but Barrett had only secured the mission houses for the company. Hapu in the various pa in the Thorndon area still refused even to consider passing over land to Wakefield.

Towards the end of April, the schooner *Ariel* slipped into this festering situation, with the much-mentioned Henry Williams on board. The forty-seven-year-old CMS missionary, with his balding head, round, wire-rimmed spectacles and formidable reputation, brought a copy of the Treaty of Waitangi to obtain signatures from Whanganui-a-Tara chiefs. He also wanted to enquire about the CMS acres. Wakefield bridled. He was already antagonistic to the treaty, to Henry Williams and to the CMS in general. The treaty was threatening the company by guaranteeing the New Zealanders possession of their land; Williams had already interfered with the Principal Agent's deals at Whanganui-a-Tara, and Wakefield knew that the CMS had opposed the theories of his brother, Gibbon, for years. Williams and Wakefield locked into what became a bitter strategic struggle, while depicting themselves in a favourable light as they reported to their superiors.

Williams finally got the signatures he wanted and, when he heard that native reserves were being set aside, waived his rights to the 60 acres (24 hectares) he disputed with Wakefield. He retained just one acre at Port Nicholson for himself and one for Reihana, his 'turn-coat' convert; but unpleasant exchanges had taken place and slanderous words were eventually written.[7] Unwisely, perhaps, Williams handed out blankets to the chiefs after they made their marks on the treaty parchment, at which Wakefield repaid the missionary's land concessions by expostulating to the New Zealand Company directors that Williams had bribed the chiefs to sign. He appeared to forget that he had given Barrett blankets for much the same reason. Williams's visit was so fraught with double or triple entendre that it is impossible to work out whether the end result came about because of Williams's or Wakefield's tact, magnanimity, obstinacy or gamesmanship, the implicated chiefs' decisions, Wakefield's colleagues' influence or advice, or Barrett's, for that matter.

However, relief was felt all round when Williams sailed away after nearly two weeks of tension, and Wakefield was free to continue his campaign to displace the people on those 60 acres at Te Aro, Pipitea, Tiakiwai and Kumutoto. As the stand-off over this area increased, Barrett's position in the middle of the two camps became increasingly uncomfortable. He was connected by marriage — if only distantly — to

Wi Tako, Wairarapa, and Moturoa, who were, in their turn, connected to Te Puni and Te Wharepouri.

The three former chiefs were discouraging, the two latter ones were encouraging, European settlement. As the Company's Agent for the Natives, one would think Tiki would be working for all iwi but, actually, he was working for the good of Ngati te Whiti and Ngati Tawhirikura hapu of Te Ati Awa, as well as for the company, the settlers, and himself.

While trying to maintain his credibility with protesting chiefs from other hapu and iwi around the harbour, Tiki had to placate them for the ravages produced by the surveyors, and persuade them that it was all in their best interests. His approach was that if they wanted the status and advantages of having white men living with them (which those protesting chiefs did not) and of lucrative trade and plenty of guns (which they did), this was the price they had to pay. As Agent for the Natives, he should have been protesting in the strongest language to his employer against this desecration. Instead, he appeared to be quite proud of the changes which his influence had already effected. It could be said that his title of Agent for the Natives was a misnomer, but this had been apparent since he had been appointed, given Wakefield's description of the job.[9] With the authority of the company, the charisma of Wakefield and the status of Rawinia bolstering him, Barrett continued to follow Wakefield's bidding to the best of his capacity.

This paid off handsomely for him. On 25 May, the Principal Agent wrote glowingly to the directors of Barrett's loyalty and success, declaring that the £100 named for him as interpreter did not seem sufficient and suggesting that, since the company had made savings in other directions, he pay him a £400 bonus. There was need for a hotel for respectable settlers, he wrote, and this would enable the seaman to establish such a hostelry.[10] The Evanses had brought with them a double-storeyed, pre-fabricated wooden house, in which Mrs Evans had intended opening a school; but she had changed her mind. The piecemeal building appeared to have the makings of the first stages of a fine hostelry, and Barrett bought it. With no thought, apparently, as to who owned the land, he had the raw materials moved next to his own home and began construction.

While they waited for this hotel, the 'gentlemen' of the colony, young jingoists with a little money and few responsibilities (though some of them were closely connected to the Directors of the New Zealand Company), soon formed a social centre in the new, rag-tag community. They christened it the Pickwick Club, and its meetings were held in the congenial ambience of one of the better local taverns in Pito-one. They made the most of their leisure moments, often on Pito-one beach, which took the place of the village green, where they played cricket or held horse races. And there was no shortage of gentlemen jockeys to ride or to lay odds on the results. Towards the end of June, they advertised in the *Gazette* a Grand Fête and Public Ball, to be held the following January, to celebrate the first anniversary of the

New Zealand Gazette & Britannia Spectator, *30 June 1840*.
ALEXANDER TURNBULL LIBRARY

**PUBLIC BALL AND FETE.**
COMMITTEE.
Mr. J. DORSET,
Major D. S. DURIE,
Mr. JAMES WATT,
Mr. GEORGE DUPPA,
Mr. F. A. MOLESWORTH,
Mr. ROBERT R. STRANG,
Mr. GEORGE WHITE,
Mr. R. BARRETT,
Dr. JOHNSTON, M.D.,
Major BAKER,
Mr. JOHN WADE,
Mr. DODERY,
Captain CHAFFERS,
Mr. JOHN THOMPSON.
Mr. J. DORSET, Secretary and Treasurer.

The Fete will take place as soon after the appropriation of the Town Sections as the Committee determine will suit public convenience.

The following list of diversions have been determined upon, and the members of the Committee named are to be addressed respecting the arrangements.

A ROWING MATCH, consisting of whale boats owned by the residents of Port Nicholson. 1st prize, £20; 2nd, £10; 3rd, £5. No race unless five boats start. Entrance fee, £2 each boat.
And a SAILING MATCH, by boats owned by residents. 1st prize, £10; 2nd, £5. Entrance, £1;—managing member, Capt. Chaffers.

A CANOE RACE, by Natives, £10 to be awarded as a prize, and a WAR DANCE;—managing member, Mr. R. Barrett.

A HURDLE RACE. 1st horse, £15; 2nd to save his stakes. Entrance fee, £3 each horse. Three horses to start. or no race;—managing member, Mr. James Watt.

A RIFLE MATCH. 1st prize, £25, to purchase a rifle; 2nd, £10, to purchase a fowling piece; 3rd, £5, to purchase a brace of pistols. Entrance fee, £1;—managing members, Messrs. Molesworth and Duppa.

WRESTLING MATCH, CLIMBING POLES, FOOT RACES, &c. £20 awarded for the same;—managing members, Messrs. Dodery and Barrett.

THE BALL.
Each subscriber of £2 is allowed to introduce the ladies of his acquaintance;—managing members, Major Baker and Messrs. Dorset and R. Strang.

All members of the Committee are empowered to receive subscriptions.

ON SALE.
BY the Undersigned, at the west end of the Beach, wholesale only, a quantity of spirits at very low prices.
T. ROSKELL.
Port Nicholson, June 30, 1840.

SHINGLES! SHINGLES! SHINGLES!
THE Undersigned has for Sale ONE HUNDRED THOUSAND VERY SUPERIOR NEW ZEALAND SHINGLES.
WILLIAM LYON.

establishment of their town. Barrett was a 'managing member' on the organising committee and donated prizes for some of the competitions.[11]

Although the Pickwick Club at Pito-one saw to it that there were fun and games offering, many of the people there chose to transfer to Thorndon. This exodus, which began at the end of May, was testing and difficult. Many of the tribal inhabitants in the now preferred area continued to show open animosity, as the Europeans scouting round for home sites trespassed on their land; and most of the physical landscape was equally unfriendly. At places between Pito-one and Thorndon the cliffs plunged straight and deep into the harbour; at others, the sea lapped in shallows, but among tumbled rocks and tree roots overhung with branches. For foot traffic, the rough, 8-kilometre, up-and-down track around the coast was passable only at low tide; but even then, there were two streams to ford at Ngauranga and Kaiwharawhara, and any villagers willing to help charged sixpence to carry travellers over.

The track was impossible for household belongings; but the sea trip was not much easier. The settlers and their luggage had been disembarked at Pito-one beach with great difficulty, from ships of 280–480 tonnes. It was even more difficult to re-embark them in the very much smaller cutters and whaleboats. Barrett and others, including Wright, provided a sea-shuttle service between Thorndon and Pito-one, and Dicky earned some money to tide him over until he opened his hotel. However, no one seemed to trust themselves or their belongings to the rather decrepit *Harriett*, and by 6 June 1840, Barrett was advertising her in the *Gazette* as being for sale again.

While this wholesale remigration gave Barrett cause for optimism, the settlers were having a bad time. Their worries about the upheaval of removal, another earthquake, another overflow of the Hutt River, and the New Zealanders' antagonism, were intensified when Acting Colonial Secretary Willoughby Shortland sailed into Port Nicholson, accompanied by the soldiers sent by Hobson to put down their treason.

The settlers, mostly peaceable landholders or indentured labourers, felt as though Hobson was treating them just like the miscreants they thought had caused Her Majesty's Government to send him to New Zealand in the first place. They treated the accusation of 'treason' as a joke; but, although they stressed their deepest loyalty to the Crown, they remained lumbered with an unbending Government Agent, who became decidedly unpopular and troops whom they had to billet and feed.

Sensing the opprobrium with which the New Zealand Company men held Shortland, Barrett may have expected Wakefield to be dismissive of the Acting Colonial Secretary, in the same manner that the Principal Agent had handled Te Rauparaha's initial high-handedness. Instead, Wakefield made placating noises to Shortland, and began planning to go north with a specially composed 'loyal address' for the Lieutenant-Governor.[12]

Shortland's presence increased the friction between the two races. The truculent chiefs at Te Aro and Pipitea gained his ear and, seeing the Crown's representative having authority over the New Zealand Company officials and sensing that Wakefield and his people lacked the support of the Queen, showed open scorn. Added to this, the previously friendly Ati Awa at Pito-one were becoming less congenial, as a result of many of the Europeans treating them with disdain or taking advantage of their helpfulness. Co-operation diminished then plummeted.

Te Wharepouri was especially personally outraged at the bandying about of new names of importance. His should be the top name, he insisted, with the Queen's second, then Hobson's; but he heard himself being put third. He sulked. The balance of power around Barrett was changing; and some of his well-being and pride, arising from those first fruitful months after the *Tory* sailed into his life, may have become shaky.

When the surveying for the town lots was finally completed, Barrett's Hotel, still only partially built, was used for the first time on 15 July as the venue for the settlers to select their lots. The finished survey plans were laid on a large table, standing in the only part of the hotel with floors, and the first lottery for 1,100 town sections began,[13] while sightseers clambered round the upper-floor skeleton of the building, or sat in the empty window frames, watching proceedings and reporting to friends outside.[14]

*Barrett's Hotel (left) was one of New Zealand's earliest pre-fabricated houses, initially intended for a school run by Mrs George Evans. Barrett's house, in the foreground, was bought for £30 to be Wellington's first library.* TASMANIAN MUSEUM & ART GALLERY

Mein Smith had divided into rectangular sections all the land from Tiakiwai pa[15] to beyond Te Aro, which sat on the best part of the waterfront, and which he had designated the commercial centre of the settlement, despite its never having been relinquished by its occupants. The names of settlers holding land-orders were put into a hat and they were allowed to choose their sections in the order they were drawn out. Mein Smith, Hanson and other involved officials stood by to answer questions, Wakefield having gone north with the loyal address.

Prime sections at Te Aro were everyone's first choice. The native reserves were included in the lottery and Mein Smith chose their sites as their lots came out of the hat. He selected Pipitea pa, on whose land Barrett's Hotel stood, as one reserve, and other Ati Awa properties nearby as others. This suited Dicky and Rawinia, who would have faced early complications if another settler had been able to choose the section on which their home and hotel sat. They apparently did not foresee the disadvantages of having their buildings on a 'native reserve'.

The lottery was exciting in prospect, but laborious and deceptive in practice. What with complaints, discrepancies and delays, it was not completed until early or mid-August. Then, of course, when the settlers moved onto their long-awaited properties, they found that they were occupied by well-established and fenced kainga dwellings. The whanau living there were outraged to have their homes, cultivations and urupa invaded, while the European lot holders were adamant that the villagers must move. Some did. Most refused to budge. Confusion merged into friction and then into confrontations.[16]

Barrett had more blankets available in the company's store, especially for these difficult situations. On one occasion, after Te Aro chiefs had been heard complaining bitterly, Barrett put some blankets into a boat, and rowed to Te Aro with Wi Tako and Brookes. The chiefs there spurned the offer of blankets; but the three men left them lying on the beach anyway, when they rowed away again, and they were eventually taken into the pa. This persuaded Wakefield that Te Aro was moving towards acceptance of white occupation; but the chiefs afterwards said they thought that the blankets were a gift for Te Wharepouri's sister, who had recently married into their hapu.[17]

Shortland had been following the action and knew that the chiefs at Te Aro insisted the land belonged to them and not to Te Wharepouri and Te Puni and the other chiefs who had negotiated with the company. When this latest unrest erupted, Shortland confronted Te Wharepouri and Te Puni with this. They confessed they had no right, then Te Puni said, 'But how could I help it, when I saw so many muskets and blankets before me?'[18]

Using 'blanket diplomacy', Barrett obtained permission for some settlers' houses, Revans's among them, to be built on ground not in tribal use. It was winter, and the villagers were seduced by the immediate comfort and warmth of wool; but they were grudging and volatile. Unwisely, one of Revans's workmen at one point made a derogatory comment against the villagers in the hearing of a Te Aro man who reacted violently and started destroying the house. A serious altercation arose. Men from both races joined in, and Barrett raced to the scene with Shortland and two officers when they heard a settler yelling, 'Murder'. European men had taken their militia guns to the fracas and tempers were at trigger point, but no blood had actually been spilt and Shortland and Barrett quickly subdued the crowd. Shortland rebuked the armed settlers, soothed the irate Te Aro inhabitants through Barrett, and arranged to sort out a solution with them all in a few days.

At this later meeting, Reihana was present as an interpreter, too. He and Barrett spoke on Shortland's behalf to the chiefs, asking them to wait until a Land Commissioner came to look into their complaints, and surprisingly persuaded the chiefs, in the meantime, to agree to 'assign over, and yield up to the Colonial Secretary all rights, titles and interest in certain lands situated in the bay, in the harbour of Port Nicholson, on which the town had been laid out by the New Zealand Company', until title to the land was decided. In return, they would keep their pa and receive compensation if it was found they had not been paid a fair price.

A crucial document, translated into Maori by Barrett, was drawn up, declaring that the government would look after the land until such time as the commissioner came. Shortland and several Te Aro chiefs signed it on 29 August 1840. Reihana testified later that both he and Tiki Parete made it clear to the chiefs that they 'were signing away their land to Mr Shortland on behalf of the Queen so that the Government might fairly compensate them for the land that had not been fairly purchased.'[19]

Barrett had had to respond to all this only a few days after his second daughter, Mary or Mereana, died, on 22 August. No cause for her death is known. Men were not expected to show their emotions, but a letter he wrote to his brother, dated 6 November 1841, shows Barrett's loneliness, homesickness and feelings for his family — both in England and in New Zealand — and grief over the loss of his daughter. He had evidently written several times before, and was 'extremely unhappy at not receiving an answer . . . and extremely anxious' to hear about his 'Family' in England, '. . . having heard nothing of them for such a number of Years. . . .' He went on to give a brief account of his activities, his employ in the

New Zealand Company, his 'Waling', his hotel, '. . . one of the most splendid building in the Place' (Port Nicholson). The letter continued:

> . . . I am now married to one of the Native Chief Women by whom I have three children the eldest whom I have named Caroline. She is now 12 Years of age — and is remarkable fine girl and is getting a blessing to me. My second named Mary was a lovely child she died to my unexpected regret about 12 months ago. My youngest named Sarah is a pretty girl and is now in good health which with the blessing of God I hope will continue . . .
> Your affectionate Brother
> R. Barrett[20]

Perhaps Mary's death contributed to the Barretts considering a return to Ngamotu. They remained in Thorndon for only a further six months; but a New Zealand Land Claims Ordinance passed by the Legislative Council in New South Wales in early August, may have influenced their decision.[21] The ordinance sought to control sales of tribal lands in New Zealand, and contained pronouncements and convoluted computations which few Pakeha and no chiefs could understand[22] and which negated all Barrett's work with Wakefield.

Sweepingly, it decreed that no payment was to be made for land in New Zealand unless it was made to the original inhabitants, and no individual sale was allowed to be more that '4 square miles'. The New South Wales Government planned to put all the New Zealand Company's 'purchases' under a microscope, as well as over 1,200 individual land claims. Wakefield and Barrett had bought 2 million acres (810,000 hectares) for the New Zealand Company, not 4 square miles, and the company had then on-sold this land to settlers, who had paid the company, not the tribes. The ordinance also prevented the chiefs from selling their own land to whomever they wished, pre-empted their ability to get the best price for their land, and denied them their sovereignty. In one fell swoop, Wakefield's purchases with Barrett at Port Nicholson and the Sounds, as well as Barrett's solo effort in Taranaki, appeared to be wiped out. There would be an immensely long time-lag before any legitimate titles could be awarded, if ever.

The Council of Colonists decided that this bombshell should also be kept from the ordinary settler, but it leaked out, turning colonists' still-hopeful dreams into a nightmare. For Te Puni and Te Wharepouri, and other Ati Awa chiefs who understood the implications, the enormity of their loss of land and mana dawned. For the chiefs at Te Aro and Pipitea, it had been something of a nightmare, anyway.

The government and the missionaries became convenient scapegoats for every difficulty and failure the company experienced. Panic swept the town. Absentee owner land-speculators were foiled and legitimate settlers, just about to get started on a block of New Zealand earth, were undone. There was a 'general stoppage of business ... workmen out of employ ... prostration of the settlement'.[23] People wandered around aimlessly, talking about cutting losses and emigrating to Chile. The *Harriett*, that old high-sterned, wall-sided cutter, was still lying unsold, derided and unwanted outside Barrett's hotel, a broomstick stuck in her deck in place of the missing mast. With his typical humour, Dicky had pamphlets printed, advertising her as available and sailing soon for Valparaiso, with a 'high poop and experienced surgeon'!

The settlers decided, however, that before they left for South America, it would be worthwhile to send a delegation to plead their case before their ultimate authority in New South Wales. Gipps, while being convinced that the brakes had to be put on the New Zealand Company, was quite aware of how demoralising this Bill would be for these innocent parties. He knew that the settlers had paid a fair price for their land, and he repaid the delegates'[24] efforts in travelling to see him by awarding them, not the New Zealand Company, 110,000 acres (44,550 hectares) around Port Nicholson.

Among various conditions specified, Gipps insisted the land must be in one block. He knew very little about tribal holdings, and had had no representations from the chiefs. The group returned home feeling reassured.

In the meantime, Wakefield had returned on 16 August, with fairly positive reactions from Hobson, too, and the nadir was passed. But where to go for country sections became an on-going problem. They were obviously not going to emerge from the steep hills around Port Nicholson, and Heaphy and a group of surveyors took off overland to Moturoa, in Taranaki, to see whether the land Barrett had bought there would be a suitable alternative.

For Barrett, however, real difficulties were accumulating. He was still doing work for Wakefield, and it was often difficult, sometimes tricky, occasionally distasteful. A dealing with the canny David Scott was all three. Wakefield gave Scott short shrift, when the latter wanted to build on the land adjoining Kumutoto pa. Scott claimed to have 'bought' it years before, and stated that Barrett could validate his claim.[25] The colonel simply said that, if Scott built, he would have it pulled down. It seems Scott persevered, and that the colonel was as good as his word. He asked Barrett and John Wade to pull down the first house Scott tried to build, then the next. This was not uplifting work for Barrett, nor could it have helped his relationship with Scott.

On top of this, by the time his hotel was finished in October, the venture had cost him much more than he could afford. News had come that Barrett's salary had been approved by the directors, but that would have taken only a drop out of the bucket of Barrett's debts. Labour was not cheap, and he may not have realized the establishment costs and the extra expenses of overheads; nor could his spirits have been raised when he actually opened for business.

By then, the economy was heading for depression. Her Majesty's Government was aware of the plight of the emigrants, as innocent pawns in the New Zealand Company's profit-seeking strategy, and was moving to effect some sort of rescue measures. But the settlers were caught in a cleft stick. Since country sections were not yet available, no one could begin farming; everyone was scrambling to make a living in trade and there was, therefore, no money from exports coming in. Householders were growing a few crops and vegetables on their still unentitled properties, but were depending increasingly on provisions grown on tribal cultivations and brought in for barter. The 'gentlemen' still had gold in their pockets, and Barrett's Hotel became the place to hold special celebration dinners and meetings; but the general public had, by now, many other 'watering holes' and competition was intense.

Despite his precarious financial situation, Barrett arranged a 'Grand Opening' for his hotel on 15 October 1840. It was reported at length two days later in *The New Zealand Gazette and Britannia Spectator* (the second name Samuel Revans had given his newspaper). Thomas Elvidge, Barrett's 'deputy', prepared a 'dinner . . . laid out with great taste', which was followed by an evening of typical fellowship for the sixteen male guests present, who each paid a guinea and a half for the privilege. The meal began with a toast to Queen Victoria, 'God Bless Her', and continued with speeches and toasts of decreasing formality and increasing hilarity. Lieutenant Best, in command of the troop of soldiers sent by Hobson, was one of the guests. Shortland was not. That unpopular governor's representative had returned to the Bay of Islands the week before. Best found the evening 'miserably dull', and the entertainment by a harpist, who was 'a monstrous beast', and 'a little deformity, who bawled yelled and screeched', well below his sophisticated tastes.[26] By the time only the fourth toast had been proposed, the assembled company was breaking into song, and by the time the eighteenth toast had been drunk, any pretence to an official function had been submerged in 'the unanimity of feeling which had been manifested'.

The eleventh toast was to Richard Barrett and family. As reported in the *Gazette*, the Chairman, who proposed it, noted that they 'were all aware of the great services rendered by him [Barrett] to the New Zealand Company; and he [the Chairman] could not neglect this opportunity of making a public acknowledgment for such service. He spoke of him as a good and honest man, one whose character and whose demeanour they were no strangers to, and of whom they ought to be proud of.' The toast was drunk to the accompaniment of loud cheers, and a Dr Taylor returned thanks, 'at Barrett's request'.[27]

The land behind the hotel was developed rapidly after the distribution of Mein Smith's sections and after Gipps's 'official' award of 110,000 acres. By the time summer came — the first in the southern hemisphere for most of the Wellington inhabitants — streets had been formed and many directors of the several New Zealand companies commemorated in their names. The sweep of roading along the beach, on which Barrett's rather grand hotel and ramshackle home stood side-by-side, became Lambton Quay; and, in November, the town became officially Wellington, named after the Duke of Wellington, a promise made *sub rosa* several years before by Edward Gibbon Wakefield in return for the Iron Duke's support, in the House of Lords, for another Gibbon Wakefield-inspired venture in South Australia. Sam Revans promptly issued the 28 November issue of his paper under the title *The New Zealand Gazette and Wellington Spectator* and thus it remained for the rest of his editorship.

Many properties on which the settlers were squatting now sported vegetable gardens, cows and budding fruit trees, but summer also brought

*Below & opposite: Wellington beach frontage, as depicted in a lithograph from a sketch by Luke Nattrass, 1841. The slopes do not show the 'network of strong, pliant creepers . . . heavy foliage of majestic trees and . . . uneven character of the ground . . . which even a goat would find impossible to tackle, which many a suburban land holder found was his lot (Wood).* ALEXANDER TURNBULL LIBRARY

*Barrett's Hotel can be seen on the foreshore, in front of the flagstaff, and Colonel Wakefields' residence on the hill behind.*

gardening difficulties, with wind, lack of rain, locusts and a 'diligent large species of caterpillar'. It also brought the local blowflies, 'fine large yellow fellows whose powers are so great that they can re-animate a piece of meat in a few minutes . . .'[28] Now servicing 2,500 souls, the business part of the town had achieved a few basic civic amenities, such as the Union Bank, which had been opened at Thorndon, while Barrett's mountainous ex-carpenter from Te Awaiti had reappeared as 'Captain' Williams, sporting an abundance of gold lace and a gold watch and chain across his substantial stomach, and joined the gang operating coastal shipping services.

There were now six other taverns at Thorndon, apart from Barrett's, and another five at Pito-one, two tiny schools, in one of which Caroline and Sarah probably started their formal education, a savings bank, an Agricultural and Commercial Club, the Wakefield Club for the 'gentlemen' of the place, lots of shops and services, a Temperance Society, and an Athenaeum and Mechanics Institute, which was looking to establish a library. Barrett, who realized that his family could live in the hotel, offered his small house as a temporary library to its committee. They accepted his asking price of £30, which made a small dent in his debts.

But the settlers were still lacking their country sections, which were

*Further along the foreshore (right) is Pipitea pa, with Wi Tako's house alongside and the cluster of New Zealand Company houses for immigrants behind.*

But the settlers were still lacking their country sections, which were supposed to be close enough to their homes for them to farm and grow exports. Both lots of land had been paid for by the settlers before they left England. After the first week of December, Wakefield made the difficult and unpopular but, apparently, only choice. Ignoring the NSW legislature's ordinance, he allocated thirty 100-acre rural sections, which Mein Smith had mapped out at the Hutt Valley, to a lucky thirty of the colonists, telling the remaining hundreds they would have to farm at Wanganui. More trouble for the future was thus put in place. Although Heaphy and the survey team had spoken enthusiastically about Moturoa when they returned, Taranaki was just too far away to be practicable.

On 12 December, the *London* anchored at the newly named Lambton harbour, carrying Frederic Alonzo Carrington, a surveyor employed by the Plymouth Company to create another settlement in New Zealand, for people from the West Country. Full of optimism, he was quite unaware that his employing company's bank had crashed while he was travelling, and that it had had to amalgamate with the New Zealand Company. Wakefield had already been informed that the Plymouth Company was to take over some of the New Zealand Company land, and he told Carrington, when he disembarked, that the area near the Sugar Loaf Islands in Taranaki would be the best place for his planned 11,000-acre town. He enthused about there being between 300,000 and 3 million acres waiting for surveyors' theodolites at Ngamotu, and expressed his view that Richard Barrett was just the man to help Carrington with the project. A year before, Dicky and Rawinia had made their major decision to quit Te Awaiti and join up with Wakefield at Port Nicholson. Now another opportunity beckoned.

*Chapter 10*

12 DECEMBER 1840–23 FEBRUARY 1841
SITING NEW PLYMOUTH

In Wellington weather at its best, Barrett went out with Wakefield and a group of others to meet Carrington, the day after the *London* anchored. The surveyor had his wife, Margaret, and three young children with him and he was not in good humour, looking at his luggage strewn haphazardly on Pito-one beach. He was aged about thirty-two, a slight man of erect bearing, with brown hair already receding from a high forehead, an aquiline nose and bushy beard. He had served fifteen years in the Ordnance Department in England, and had a reputation as a respected, conscientious, skilful surveyor and draughtsman.

Now Chief Surveyor for the Plymouth Company, he carried detailed instructions[1] and a contract giving him equipment, tents and rations, £300 a year, four pence an acre for all land surveyed — as soon as it was sold — plus a 1 per cent commission on the purchase price of each of the town or country properties. In theory, if he worked hard, he could make a small fortune quickly; but he was going to make his own decision about where he did so and who helped him. He waited six days before inviting Barrett to join his team.

It should be remembered that Wakefield had not set foot in Taranaki, and of the men who signed the Ngamotu Deed in February 1840, Barrett had the strongest motive to put it forward as the site for the Plymouth Company. For years he had been a nomad. Te Awaiti had promised much, but overfishing by foreign vessels had put a damper on that prospect of security. A new settlement at Ngamotu offered him fertile soil, whales, a more temperate climate than Wellington or Te Awaiti, more trading opportunities with a prospective European population, and a return to his ready-made extended family. Yet he requested a week to make up his mind. Why the hesitation? Could the birth of his two sons this month have caused this procrastination? There are no answers.

Like Mein Smith before him, Carrington had Plymouth Company settlers on his heels. While awaiting Barrett's decision, he steadily and meticulously stocked up for the coming expedition, Wakefield having made available both the company's storehouse and the 227-tonne barque *Brougham* for Plymouth Company use. Christmas followed on, hot and calm, and two days later the three men met. Barrett came out with his terms. He agreed to go to Taranaki, but only on condition that he was paid £100 for the term of the trip. Carrington, careful of Plymouth Company funds, was unwilling; but Wakefield pressed him: 'Mr Carrington, you had better make up your mind to the expense. You must look to the future result of the company & the great advantage of Mr. Barrett's knowledge of the country & language. I wish to make a point of writing to your Directors & bearing you out in being perfectly right in taking Mr. B . . .'[2]

Carrington could do nothing but give way, and Barrett was soon engaging in just as much hectic preparation as his new employer. He made arrangements for a group of Ati Awa family and friends, who were keen to come to Ngamotu with him and Rawinia, to travel on the *Brougham*, bought the only horse he could find for sale, which had a foal at its side, put a manager in his hotel, accepted appointment as *Gazette* agent in Taranaki, and advertised in the community's newspaper:

BARRETT'S HOTEL

R. Barrett begs to inform the Colonists of Port Nicholson that the above Hotel is now open, and trusts by affording every accommodation to his guests, to merit a continuance of the patronage already bestowed upon him.

Wines, spirits, ale, and porter, of the first quality.

An ordinary daily at two o'clock. Cold collations always ready. Good beds.[3]

*NEW ZEALAND GAZETTE & WELLINGTON SPECTATOR*, 26 DEC. 1840

*Carrington did not relate well with Barrett initially, and referred to the native New Zealanders as 'horrid savages'. However, by the time he returned to England, he and Dicky were much closer and Carrington was much more understanding of Maori needs. In 1857, he came with his wife and family to live in New Plymouth and died in 1901, known as 'The Father of New Plymouth' (DNZB p.73).* TARANAKI MUSEUM

Notwithstanding, he arranged to take much of his hotel equipment and stocks of wines and spirits with him, since he had plans for another advertisement about a Taranaki Barrett's Hotel. He then began loading these into the *Brougham*, plus household belongings and whaling equipment, confirming that he and Rawinia were intending to move out permanently. They would get a free passage to Ngamotu, even if Carrington did not choose it for his town.

He was also keeping a friendly eye on Carrington. He pointed out to the surveyor that stores were unavailable around the country, and mentioned certain goods as being very attractive bargaining tools. Carrington, as a result, increased his cargo by half a keg of tobacco and three hundredweight of biscuits, as well as more flour, rice, brandy and general stores from the Wade brothers. However, he and his four-man crew were caught by a sudden wind change as they tried to get these provisions on board the *Brougham*. They struggled against the harbour's gigantic waves for two and a half hours in their row-boat before Barrett came to the rescue and towed them to the ship with one of his whaleboats.

Another indication of Barrett's intention to leave permanently was his absence from a meeting on 4 January 1841, ostensibly close to his heart and interests, to form a Flax Association in Wellington. He was much too busy and Carrington was needling him to be ready. Finally Barrett packed his family and the horse and foal onto the *Brougham*, and she weighed anchor early on the morning of 7th or 8th of January, beating out of the harbour into a south-east wind, then sailing fair for Taranaki.

The mild conditions gave Barrett an easy thirty-six-hour trip as pilot, and he slipped easily into his leg-pulling, story-telling role. Interpreting between Carrington's party and his own group of Ati Awa, he entertained with accounts of his experiences and escapes, ranging from highly exaggerated to completely fictional, for the benefit of the uninitiated. 'More than once, he tells us, he has been tied to the stake preparatory to becoming a meal for one of the chiefs', one surveying assistants wrote.[4]

The sea was calm when they reached Ngamotu at 5.10 p.m. the following afternoon and fired four guns as they anchored. Within minutes of the reverberations dying away canoes were at the *Brougham*'s sides. Figures scrambled on board and there was an emotional repeat of the reunion on the *Tory* fourteen months previously, but on a larger scale. Then Barrett off-loaded a whaleboat and he, Rawinia, Caroline, Sarah and their relations, along with much of the gear that Barrett had brought with him, were packed into that craft and the canoes. Heavily overloaded, they paddled to shore, heading for the small collection of huts near the base of Paritutu.

There were other habitations to be seen. On the north bank of Waitapu stream, a 13 x 6.5 metre mission house now stood lonely and unoccupied on Meurant's purchase[5] and, a kilometre along the beach towards the kainga, three very large reed houses faced the sea. Even the smallest was around 30 metres in length, and 6 metres wide, stoutly constructed with flax-bound reeds. The surveyors who had travelled to Moturoa with

*Ngamotu and the Sugar Loaf Islands (now known as the Sugarloaves) — dominated by Paritutu on the left and Moturoa, with several small huts clinging to its sides. Barrett's three houses can just be discerned at the top of the white cliffs, next door to Ngamotu kainga on the steep promontory.* ALEXANDER TURNBULL LIBRARY

Heaphy had brought back tales about this accommodation, which, Barrett claimed, the locals had built for immigrants at his request.[6] Barrett returned to the ship later on his own, to stay with the horse and her foal overnight.

He took Carrington to the beach the next morning, at the start of his apparent campaign to persuade the surveyor to confirm Wakefield's suggestion of Taranaki as the place for the new settlement. The surveyor was greeted by handshakes from the fifty or sixty people who crowded round, and repeated cries of 'Haere mai! Haere mai!' When Carrington asked what this meant, Barrett said that they wanted white people to come; then went on to intimate to the surveyor that the clustering, welcoming crowd were talking about little else but the expected arrival of Europeans coming to live with them.

The mare and her foal were then brought ashore, onto black sand hot enough to fry eggs on, but Barrett mounted the horse and galloped off her energy, pent-up after two or three days' confinement. The villagers scattered in fright from the strange beast; but by mid-morning one tribesman asked to ride her and Barrett gave him the reins. The young man vaulted onto her back with casual confidence, gave her a wallop, and he, too, galloped off, with hollers and waves and natural balance. He returned half an hour later grinning and triumphant, the horse in a real lather after galloping along the shoreline and back. Barrett was not fazed.

A tour of the island of Moturoa came next, and from the top of it Barrett showed Carrington all the land he had 'bought', right up the coastline to beyond Parininihi. Carrington climbed up to the whata, via the access trunk with its notched footholds, and examined the huts and store caves and the Barretts' rock-shelf home. He measured the sea waters off Ngamotu as two and a half fathoms, and talked enthusiastically about the feasibility of building a furlong of breakwater out from the beach to the middle island.

On 11 January, the two of them, plus four surveying assistants — Aubrey, Baines, Rogan and Nesbitt — set out at 9 a.m. to examine Waitara in Barrett's whaleboat, rowed by a seamen crew. It was extremely hot, and it took four and a half hours to reach the river's mouth. Carrington noted the tidal harbour it created and commented on its exposure to the northwest winds, but Barrett assured him that they occurred infrequently and lasted only a few hours. The seamen resumed rowing, crossed the sand bar, then went up the river for nearly 5 kilometres, before landing and cutting their way through the 2-metre-high fern to explore. Carrington was delighted with the rich soil and the neat and healthy village cultivations they found, saying they were in 'the garden of the Pacific'; but he was less than impressed by the few Ati Awa tending the gardens 'in the greatest state of wretchedness'.[7] The boat load slept that night in the open, and even the sandflies and mosquitoes did not upset Carrington's enthusiasm for the place as first-rate agricultural land, but their itches did start them on their way back to the ship early in the morning.

When the party reached the *Brougham*, after a long, hard row, Carrington — in almost as much of a hurry as Wakefield had been in buying land for Wellington — ordered the anchor up and pressed on that afternoon. He had decided against the Sugar Loaves area, he said, because of the absence of a decent harbour. He wanted to look at suggested places in the South Island. Barrett left his family at Ngamotu and went with him.

Before Carrington left Wellington, Wakefield had agreed that the New Zealand Company would allow the Plymouth Company to use Port Hardy at D'Urville Island as the anchorage for any town at Taranaki. Carrington headed there now. He saw a great harbour indeed, when they anchored at 6 p.m. on 15 January, but a complete absence of level land for settlement; however, he noted that 'Mr. Barrett thinks the quantity of crooked timber at Port Hardy will become very valuable for Ship-building & that it has advantages over many ports for repairing large Ships. You can also ballast with little trouble.'[8]

Barrett appears now to have manipulated the rest of the exploration on three occasions, as they steered the *Brougham* from place to place. Firstly, after sailing along Blind Bay, when they next anchored he took them to explore the mainland opposite Adele Island and on to Motueka. It consisted of 800–1,200 acres (324–486 hectares) of level land and valuable timber trees, but much of it was swampy and obviously prone to flooding. Here they received a warm welcome from three Maori who boarded the ship. 'E Tiki,' they called out and invited everyone to a feast of fish and potatoes with their whanau. The ship's company was captivated by these people's kindness and attention, but Carrington was not so taken by the terrain he saw that day.

On the 20th, Barrett took Carrington and Baines on an exploration they would not wish to repeat, this time of obviously quite unsuitable ground. They went ashore, through many 'sand spits and shoals which made landing difficult', at a desolate spot. There was no tree to be seen for miles, only stunted grass, fern and scattered bushes, and they were lucky not to be 'swampt' returning to the ship in a nasty swell. That Barrett may have deliberately chosen to show Carrington this barren land, could be the second example of his tactics. Eight kilometres further on, and eight out, the surveyor wanted to venture in again, but Barrett managed to steer him away, saying, 'Too far to venture in a whaleboat ... through so much sea and wind.'[9] This, a possible third example of his strategy, was remembered and commented on unfavourably two years later, when Nelson and its harbour were eventually established at that spot.

Everywhere now were 'mountains to the sea' all the way back to D'Urville Island, after which, the land being almost totally covered in mist, Carrington had the *Brougham* bear up for Queen Charlotte Sound. On 21 January, a 'perfect gale' from the north-west played into Barrett's hands in another way. The high wind allowed him to pilot the *Brougham* past Ship's Cove, which was on the mainland, and to anchor it instead in East Bay, which was, of course, on Arapaoa Island. When they anchored at 1.30 p.m. Barrett had five guns fired, a whaleboat came out to collect him in the wet, cold and windy conditions, and he was able to go ashore to attend to business at his old whaling station.

After he returned to the ship the next morning, Carrington took the *Brougham* round briefly, to have a look at Te Awaiti, but did not linger, pushing on to anchor at Lambton Harbour in the early evening of 24 January. The whole jaunt was over just in time for them to experience the regatta, hurdle races and 'Popular Ball', the last among the long-planned festivities celebrating the town of Wellington's first birthday, for which Barrett had offered to donate prizes. They were also in time for Carrington to greet his brother, Octavius, who sailed in on the *Slains Castle*.

Two days later, as a result of Barrett's Arapaoa Island foray, two whaleboats and much equipment from his whaling station appeared on the Wellington waterfront. On the same day, Carrington informed Wakefield that, despite his initial misgivings, he had decided that New Plymouth — the proposed town already had a name — should be at 'Taranake' (*sic*), but on condition that Port Hardy be made available to the Plymouth Company. Its superb anchorage was close enough to the Sugar Loaves for it to be used as a provisional harbour. None of the sites he had explored with Barrett as pilot, he intimated, had had as many advantages as the Taranaki coast. Barrett's apparent strategy had been successful.

Wakefield responded instantly. He urged Carrington to return immediately to begin surveying at the Taranaki site, making the *Brougham* available again and expediting its departure by saying that he and Ockey

(Octavius Carrington) would see about the stores, insurance, and any other matters. The colonel had not even seen those Taranaki lands, but he had possessed himself of its soil through Barrett, and was now intent on sending occupants as soon as possible. Spurred on by the colonel's enthusiasm, Carrington began loading the *Brougham* between bouts of dreadful weather. Barrett meanwhile turned his attention to the vast number of details arising from moving his household back to Ngamotu. One of them was making his will:

> This is the Last Will and Testament of me Richard Barrett of Wellington Port Nicholson Licensed Victualler, being of sound mind, memory and understanding. Whereas I am possessor of Several Tracts of Land and hereditaments in New Zealand and elsewhere, more particularly of an allotment of Land Situate at Wellington aforesaid described as Number 514 in the Surveyors Generals plan of the New Zealand Land Company's first and principal Settlement, in right of Wakaiwa my Wife a Moari Woman upon which I have Built a good and Strong wooden Massuage or Inn Known by the Name of Barretts Hotel . . .[10]

It ran to seven wordy, legal pages, leaving one third of his estate to Rawinia (Wakaiwa), with the proviso that, if she remarried, her next husband, or husbands, should never gain control of her inheritance. The rest of the estate was to go to his children, with the usual provisos about children's inheritance in the event of their marriage, or their predeceasing their children, etc. Unusually, he bequeathed to his children equally, at a time when sons by custom inherited much more than daughters. He made provision for any sons — after all, Rawinia was only in her later twenties; but it raises the question of whether he knew about his two supposed sons (Henare and Edward) and was giving them rights, despite his having acquired property through his wife, who was not their mother. Also, although the bulk of his supposed estate, apart from chattels, came to him through his wife's connections, he precluded any husbands of his wife or daughters having any control over their inheritance! He named David Scott as one of his executors (along with one William Vetruvius Brewer), somewhat surprisingly, after Barrett's actions against the former. Thomas Waters and James Smith, who were to feature prominently in Barrett's life before long, were two of the three witnesses.

Afterwards, he and Carrington continued packing their supplies and equipment and making all their other arrangements to leave Wellington; and, on Tuesday 2 February, the surveyor gave Barrett the promised £100. The Principal Agent, however, was finding himself responsible for a multitude of unexpected expenses, and, on the same day, Wakefield told Barrett that his name had been removed from the company payroll. This would have been a major blow to Barrett, but understandable, since he was moving into the New Plymouth Company area and out of easy contact with William Wakefield. With funds in his pockets, Barrett used Revans's pages for a succession of basic 3/6$^d$ advertisements. Mr U. Hunt took over a lease as manager of the Wellington Barrett's Hotel, and Barrett tested the waters by inserting another advertisement for a different hotel.

> BARRETT'S HOTEL AND STORE, TARANAKI.
> ... he intends opening on the 1st of March next, the above Hotel and General store, with a well-selected and choice stock of wines, spirits, ale and porter, and general assortment of goods suitable for settlers, at very moderate prices; and trusts by attention to the convenience and comfort of his guests to merit a share of their favours.
> N.B. — Good beds and accommodation for travellers, and private apartments for families.[11]

He also left advertisements, to be inserted in future issues, advising he would arrange for mail to be delivered between New Plymouth and Wanganui, and announcing his appointment as the newspaper's agent at the Plymouth Company's settlement at Taranaki.

During the first week of February, the *Brougham* settled down in the water as Barrett's boats, gear and whaling appliances, and everyone's furniture, animals and plants, were put on board. Carrington, at the last minute, arranged for prefabricated parts of a wooden house to be taken, and then had trouble finding a place on deck, amongst all the clutter, for the goat Wakefield gave him.

By Sunday 7 February, Carrington was again telling Barrett firmly that he wanted to sail the next morning. Carrington omitted to consider that most of his own possessions had been packed since he had sailed from England, while Barrett, on the other hand, was having to pack up, or make arrangements for, everything from his previous life and livelihood, that had not been flung onto the *Brougham* during its previous trip to Ngamotu. Co-operative as always, Barrett said he would be ready first thing; and, sure enough, on the Monday, the wind having moderated and the weather look-

*Barrett advertised several times in* The New Zealand Gazette & Wellington Spectator. *Charges for advertisements were: six lines and under, 3/6$^d$ for the first, and 1/- for each additional, insertion; six to ten lines, 5/- and 1/6$^d$; over ten lines, similar increases. Subscribers to the* Gazette *paid 40/- per annum or 1/- per issue.* ALEXANDER TURNBULL LIBRARY

---

**BARRETT'S HOTEL, WELLINGTON.**

U. HUNT begs to announce to his friends and the public, that he has succeeded Mr. R. Barrett in the above establishment, and trusts that by attention to secure a liberal share of support from those who may favour him with their patronage.

Wines, spirits, ale, and porter, of the best qualities.

An ordinary at 2 o'clock daily. Cold collations always ready. Public dinners provided.

Good beds, and private apartments for the use of families.

Wellington, Feb. 19, 1841.

---

**BARRETT'S HOTEL AND STORE, TARANAKI.**

RICHARD BARRETT begs to inform the inhabitants of Port Nicholson, and the adjoining Settlements, that he intends opening on the 1st of March next, the above Hotel and General Store, with a well-selected and choice stock of wines, spirits, ale and porter, and general assortment of goods suitable for settlers, at very moderate prices; and trusts by attention to the convenience and comfort of his guests to merit a share of their favors.

N.B.—Good beds and accommodation for travellers, and private apartments for families.

---

A LARGE roomy substantial Warre to be let by the week, month, or year. For particulars, apply to Mr. W. H. Bottomley, at W. V. Brewer, Esq., barrister-at-law.

---

TO BE SOLD, or exchanged for mares, one bay horse, warranted the best draught horse in the Colony, and one grey horse.

To be sold, or exchanged for cows, 5 superior working bullocks.

J. C. CRAWFORD.

---

**WANGANUI.**

LAND AGENCIES.—The Undersigned having last month examined the sections about to be chosen at Wanganui in March next, offers his services as Agent there, being about to proceed next week, (per the "Elizabeth,") as a resident settler. Letters addressed to the care of Messrs. Ridgways, Guyton, and Earp, will be forwarded.

JOHN NIXON.

Wellington, February 12, 1841.

---

LAND AGENCIES.—TO LANDHOLDERS.—The Undersigned beg to offer themselves to their friends and the public as Agents for the disposal of Town and Country Lands, and trust their colonial experience may entitle them to a share of the public patronage.

They will be happy to arrange for and superintend the surveys and sub-divisions of the land. Liberal advances will be made on lands placed in their hands for sale. They will also be happy to assist proprietors in leasing their lands.

JOHN and GEO. WADE.

Wellington, Oct. 7, 1840.

ing fair, Colonel Wakefield and others came to the waterfront and farewelled them all. The ship's crew fired off five guns as she left, which were answered by the same number from the shore, and she moved gently down Wellington harbour at a quarter past five. Barrett did not discover until later that, in the rush, vital whaling tools had been left behind.

On the *Brougham* were the Carringtons, their three children, and Frederic's brother Octavius (Ockey), Aubrey, Baines, Nesbitt and Rogan and another five or so surveying assistants and two of their wives, and Charles Nairn and his sons. Also on board were passengers Spencer and Jervis, taking merchandise to open a store, one other settler accompanied by three farm labourers, and a few whalers, their Maori wives and children. Wright, Bundy, Bosworth and 'Black' Lee may have been among them.[12] They are reputed to have stayed close to Barrett throughout his years in New Zealand, and with him were the original settlers of New Plymouth.

Once out of shelter, they were hit by a violent nor'wester and, when they were still only off Kapiti in the morning, the captain decided to shelter in Ship's Cove until the winds were more favourable. This suited Barrett. Perhaps he even suggested it. He still had provisions at Te Awaiti, and friends and relations of Rawinia's keen to return to Ngamotu. He left the *Brougham* on the afternoon of 9 February and made his way to the whaling station, returning the next morning with two more whaleboats full of people and stores, presumably with the permission of the *Brougham*'s captain and Carrington. The whaleboats, still full of stores, would have travelled the short distance remaining stacked one on top of the other and lashed down to the deck. The people, used to discomforts, would have been squeezed in somewhere.

Two days later, on Friday 12 February, the *Brougham* anchored just over a kilometre north-east of the centre Sugar Loaf. For the second time, Dicky and Rawinia became founding members of a New Zealand town; and, for the fourth time since arriving in New Zealand, Barrett began the process of settling down. On this occasion, he had left behind 2,500 people in the town of Wellington, to come to Ngamotu with its sixty- or so inhabitants, overgrown terrain and widely scattered cultivations.

*Barrett's three houses were described as being nearly equal in strength and durability to the prefabricated wooden ones brought from England at great expense.* TARANAKI MUSEUM

Carrington was intending to sail on, with his party and passengers, to where the town was to be sited at Waitara, but he agreed to remain anchored off Ngamotu until Barrett had disembarked. This involved loading all his provisions and gear into the boats and canoes while they bumped up and down alongside the barque, paddling them to the beach, unloading again and lugging them up the slope to the three houses, then repeating the process for the whalers' belongings. Barrett warned the European paddlers about Tokomapuna, a long line of rocks, running out from the beach towards the north and treacherously just covered at high tide. The settlers promptly named it another Barrett's Reef.

Night fell before they had finished, and while the survey party and other passengers bunked down on board, the whaling team settled themselves into one of the three large huts. According to one source, this became their permanent home, where they and their Maori wives evolved dormitory-type accommodation, with 'spaces at the sides of the whare being partially staked off, and beds of fern laid down ... In each of these partitions were placed the occupant's blankets, pannikins, spoons, knives and forks, &c. Ranged down the centre of the whare, on the ground, were iron pots, rations of meal, potatoes, and a bucket of rum.'[12]

The next day was Saturday 13 February, and Carrington despatched Ockey and a team in the ship's boat, surrounded by provisions, tents and tools. They were to build sheds for stores and accommodation at Waitara, and he and the *Brougham* would follow them when Barrett had finished unloading. Nairn returned the next day, however, to report that three in the boat had nearly drowned in the 'truly awful' surf on the Waitara bar. Carrington went to investigate, and returned dolefully demoralized on the 16th. It had been the availability of this harbour that had decided him to site the settlement here. He reconsidered. Wakefield, he remembered, had actually recommended Moturoa for the site of the town. The superb soil would still be not far away, and Barrett had assured him that the winds on that coast were fair for nine months of the year. Ships would be able to unload in the roadstead or shelter at Port Hardy. With time at a premium, Carrington ordered the Plymouth Company cargo to be unloaded on the beach. Barrett was going to have his town next door.

From dawn on 17 February, company stores and equipment were discharged onto a raft and into a boat; then, as the breeze increased, another company boat was put into service, and finally, as the waves built up, Barrett's boats had to help them out. They continued to be used on and off over the next four days of mixed summer weather, until the *Brougham* was sitting once again high and unstable in the water. Ballast was needed. Before the ship left, Ockey settled all the accounts with the captain, including paying for one extra bag each of rice and sugar from the ship's stores. No payment was offered to Barrett. At 10.30 a.m. on 23 February, the captain fired a seven-gun salute and sailed the almost empty ship down the coast, looking for something weighty to replace the bricks and shipload of cargo lying on the beach. The Barretts were home, and the settlement of New Plymouth was left to its own devices.

*Chapter 11*

23 February–11 August 1841
A WHALER AGAIN IN HUNGRY NEW PLYMOUTH

On the slope up from where Otaka pa had stood, Barrett built an unpretentious reed house for his family, neither as solid nor as large as their Te Awaiti home, nor even their Wellington one,[2] but with a wide view out to sea for their quarry and away from whaling miasma. Jervis, the would-be storekeeper, later wrote that Barrett at the time was '. . . a thick-set, podgy little fellow, with a good, natural honest countenance, and yet chief amongst his rough, burly comrades . . .'[1] Barrett was now about thirty-four. Rawinia would have been approaching thirty and Caroline and Sarah eleven and six.

Rawinia's mother and father lived nearby and Hongihongi lagoon, for all their ablutions, was just below to their left. From the house, the incline ran gently down to the beach, where the whaleboats would lie, and where they would construct the station; but Barrett was well aware that there were serious gaps in the supply of lines and ropes, and not enough whaleboats to form an effective hunting pack. He would send urgently for those he had left in Wellington.

The Wesleyan mission house next door now had a small companion hut, housing English neighbours. The Rev. Mr Creed and his wife had arrived to set up a mission at Ngamotu just one month previously, helped temporarily by the Rev. Mr Wallis. They were already well spoken of by the locals. When the Barretts arrived, however, they were absent, taking the Word to southern villages. The mission station's north-eastern boundary abutted Otaikokako beach, and the north-western boundary was next to where the palisades of Otaka pa had stood.

The region was just as they had left it fourteen months before, head-high in impenetrable fern and scrub. Exhaustive clearing had to be done before they could even think of planting for the future. Only 10–15 acres (4–6 hectares) of widely scattered lots were under cultivation by the few remaining inhabitants, who could spare only minimum effort for Tiki. Fortunately, he had his whaling team to help.

Fire-breaks therefore became Barrett's next priority, in readiness for when he lit the scrub before tilling the soil, and for safety from the surveyors' fires, too. They were already cutting wide swathes around their buildings, before burning the scrub and laying out their survey lines. Everything was tinder dry after the summer. Frederic and Ockey Carrington and Baines were standing near some untorched bushes which went up with a whoosh in a wind-change one day, and saved their lives only by making a quick dash across some smouldering red-hot ashes.

Carrington had decided to site the town 3 kilometres away from the Barretts' place, between the Huatoki and Henui streams. He wanted no tangle with the Wesleyan Church's hundred acres, nor with the area occupied by the Barretts, nor the kainga at Ngamotu. There were other factors too, he intimated: water was plentiful, the forest for house construction and firewood would be closer, they would avoid the sand, which would blow in the dry weather, and the town dwellers would be well away from Barrett's planned whaling station, 'which would be a horrible nuisance'. This still gave him plenty of space for a settlement of 2,200 quarter-acre town sections that his contract required. He was envisaging the 209 50-acre (20-hectare) suburban sections and 1,150 50-acre rural sections further north, towards Waitara.

Within days, the town site became a vast expanse of tangled, charred remains. Te Ati Awa, drifting home in small groups since news of Tiki's negotiations for Taranaki land had flitted round the country, were not at all pleased with the condition of their tribal land. They were surly and antagonistic towards the surveyors; and upset with their relations, who had alienated their land and kept the goods they had been given for it.[3] Tiki was often reminded about his promise of double-barrelled guns.

In early March, the *Jewess* anchored off-shore, on its way to Port Nicholson, and its captain, George Moore, offered them more provisions in the form of pigs and potatoes from Mokau. Barrett took advantage of this visit to send an urgent request that his whaling gear in Wellington be forwarded to him on Moore's next trip. Prompted by Barrett, Carrington wrote to Wakefield, asking equally pressingly for the promised firearms.

> Taranaki
> Sugar Loaves
> 8 March 1841
> My dear Sir
> ... Mr Barrett tells me, when this place was purchased from the Natives, that an agreement was made to make some few presents to the Chiefs, when the <u>white people</u> came to settle here, which presents they are now anxiously inquiring about.
> I shall therefore feel particularly obliged if you will be so kind as to send me something in the way of <u>trade</u> to satisfy these people — I am told 'double barrel'd' guns were promised them ... I fear from what I have seen, unless the promise is fulfilled we may have some little trouble ...[4]

Indications of this trouble were soon apparent. The day after the *Jewess* left, Barrett was called in to help when two Waikato tribesmen 'danced and flourished their tomahawks' at Carrington, protesting about a shed he was having erected on the banks of the Huatoki. The surveyor felt very threatened and inadequate, and 'returned home ... for the purpose of seeing Mr. Barrett ...', whom he needed as interpreter. Barrett tramped to the riverbank the next day and discovered the trouble had arisen because the designated spot was temporarily tapu. With some good-humoured talking, patient bargaining, and a blanket or two, a surveyors' hut was finally allowed to be constructed in another, just as suitable place. In Carrington's mind, the event became the time that 'the Waikatos' threatened to cook him and Barrett.[5]

The latter's stories on the *Brougham* had recently been given credence. Barrett had a simple store-whare at Ngamotu, where his men could buy

*Charles Creed, the first Wesleyan minister at the mission station next door to Barrett's Ngamotu home. He married Dicky and Rawinia and baptised their daughters Caroline and Sarah.* HOCKEN LIBRARY

their provisions on credit, though it hardly lived up to the 'Hotel and Store' he had advertised in the *Gazette*. Jervis and Spencer had decided to set up their slightly better-equipped shop not far away; but as they dug foundations they unearthed quantities of human bones, including skulls obviously smashed by patu or tomahawks. The locals took them away for burial, after payment of tobacco utu, but not until they themselves had given graphic descriptions of 'roasted and kopa-maoried Waikatos'.[6]

The Creeds and Mr Wallis came home soon after, during the third week in March 1841 and were extremely gratified to conduct the first Divine Service with whites among the congregation on Sunday 21 March. On the following Sunday, amid a gathering of whanau and Europeans, Tiki and Rawinia were married, after about thirteen years together. Caroline and Sarah were baptized at the same time.

For Dicky, a Christian Englishman, the marriage may have set a seal of respectability on their union; but there may have been an ulterior motive. He had not married during his year in Wellington, nor earlier, when Bumby and Hobbs were at Te Awaiti. The assets in his newly made will came from Wakaiwa and her tribal connections. He could have been tidying up loose ends. Carrington commented, just over two weeks later, that several of the white men in New Plymouth had married their long-time partners during the past week, and he reckoned they may have done this to achieve land titles.

Rawinia was an industrious partner, and had an extensive vegetable garden where she planted quite a variety of produce, helped probably by Caroline and Sarah, just like many of the pioneer wives and children. As they grew older, the girls as a matter of course would have worked in the garden, cut vegetables, collected eggs or caught fish, and been often involved in a number of regular hapu activities associated with food growing and gathering. In addition, they would have learned weaving and other female skills. Mrs Creed, 'E Mata' to Te Ati Awa, soon began offering lessons to the children of the whalers and the tribe, which Caroline and Sarah would have attended. They both grew up as literate young women and married brothers from the same settler family.[7]

Immediately after the wedding, Barrett took Nairn and another man 5 or 6 miles inland to look for his 'Australian' peach trees, left unattended since the heke, and so missed the arrival of the first Plymouth Company settlers, in the 312-tonne barque *William Bryan*, on Tuesday 30 March 1841. The three men returned that Tuesday night, their arms full of peaches and cuttings, to find Carrington with two unknown Englishmen inspecting the site for the new town. The two strangers were George Cutfield, the temporary agent for the Plymouth Company, and Henry Weekes, the company's doctor. In Barrett's absence, Carrington had saluted the ship with shots from Dicky's cannon[8] and used his whaleboat and crew to go out to welcome the barque himself.

On the spot, Cutfield asked Barrett whether he would make his men and boats available to complete the unloading. All the immigrants and their livestock had been landed but the company labourers would need help with the tons of personal possessions, tents and prefabricated houses which remained in the *William Bryan*'s hold. Barrett agreed to the request, with Carrington, Weekes and Nairn as witnesses; yet, as the new settlement grew and the need for boats and services increased, Barrett's boats were used for any jobs as of right and, it seems, without permission, thanks or remuneration.

Continuing on home that night, the three peach-seekers discovered a small instant settlement waiting for them. Some immigrants had found their own tents in the ship's hold and erected them on the beach, and well over a hundred others had joined the Carringtons and the few white whalers in the three large whare. Most of the surveying assistants and labourers, who had still been living in them, had had to move to the storehouse on the Huatoki stream.[9] The fifty or so resident Ati Awa adults and children had welcomed these travellers from England, as custom decreed, with handshakes, potatoes and watermelon. Not so welcoming were an unexpected plague of locusts, hopping and flying everywhere, and rats, which quickly discovered the provisions lying under tarpaulins on the beach and in the stores. The locusts passed on. The rats remained, along with millions of sandflies, as constant irritants and destructive pests.

The next day, before his boats began work, Barrett moved the *Brougham* to the new anchorage he had certified, in a good, deep channel between Moturoa and Motumahanga, beginning his stint as the settlement's unofficial, unpaid harbourmaster over the next few months.[10] Then, in blessedly fine weather, the disembarkation continued unabated for a week, the goods piling up in considerable chaos 'on a bank just above high water mark, under Mr. Barrett's houses'.[11]

Barrett had no responsibilities for these belongings. The settlers themselves, both cabin class and assisted labourers, had to carry and drag their luggage along the beach and through the blackened, burnt-out wasteland for over 3 kilometres, fording the Huatoki, to two temporary housing sites on the slopes of Mount Eliot,[12] set aside by Carrington to be 'squatted' on until the town lots were ready.

Therefore, in a week or so after the *William Bryan*'s arrival, New Plymouth was spread out and already divided into factions in different spots. There was the native village, containing also Barrett's home and the three large houses for immigrants, whalers and labourers, at Ngamotu;

*George Cutfield, Agent for the insolvent Plymouth Company for only a short time, impressed Te Ati Awa by a display of fireworks on Saturday 19 June 1841, two and a half months after his arrival, and at times made life difficult for Barrett.* TARANAKI MUSEUM

the missionaries on their land next door; some of Carrington's overworked surveyors' assistants and labourers 3 kilometres away, on one area of about 5 town acres, and others near them in the storehouse on the Huatoki; the dissatisfied, impatient settlers, also divided into upper and lower social strata, squatting on the slopes of Mount Eliot; and the tribespeople, increasingly bewildered and hostile, scattered around their 30,000 acres — although one colonist wrote home that Dicky Barrett had 'them quite under control'.[13] All groups of both races were uneasy, living not only with unaccustomed neighbours, but also with constant rumours of attack from the north, and fighting between Wanganui and Tuwharetoa in the south. But, spread out though they were at New Plymouth, their futures were now inextricably linked.

The relationship between Cutfield and Carrington quickly became strained and Barrett became involved. In charge of this batch of Plymouth Company immigrants, Cutfield had immediately assumed complete authority, and had begun using Carrington's surveying labourers to build a bridge over the Huatoki stream and a storehouse. Carrington, in high dudgeon, asked Cutfield to have them first erect his prefabricated house. He was living in a raupo hut with no floors, doors or windows, crammed in with some labourers and their luggage, all the company stores, and his wife and three unruly, beloved children. They had to wear greatcoats and cloaks inside when it was at all cold, and he was having to come in and draw plans by lamplight at night after being out with his instruments every hour of daylight.

Cutfield witheringly intimated that the surveyor's house should be a 'private' job, and not one for company workmen. For all that, the next morning at low tide Barrett found Cutfield — in what looked like a fit of pique — having the pre-packaged house rolled down Otaikokako beach, for transport along the bay to the town, even though the north-westerly had got up late the day before. Barrett was perturbed. There was no way the raft for its transport could be launched into the heavy waves pounding the beach; and the wooden crate holding the house would be broken up when the tide came in, if it was left where Cutfield had taken it. Carrington turned up and was furious. Barrett said nothing until Cutfield appealed to him, then 'amiably pointed out the folly of what he was doing'. Cutfield stormed off and Carrington angrily ordered the package to be put back a safe distance above high-water mark, where he secured it.[14]

Carrington continued with his town planning, and Barrett pondered his plans for the area behind Ngamotu, between his home and the great forest. Through deceit or self-delusion, he had exaggerated the East Bay Deed's informal codicil into some thousands of acres, on which he proposed establishing a farm. This was the time of year for cutting and burning bush, in preparation for breaking in the land. However, Carrington heard about Barrett's intentions, and after dinner on 13 April, with Baines in on the conversation, he tackled Barrett, letting him know that that land, as well as Waitara, was earmarked for Plymouth Company rural sections.

Barrett's usually smooth hackles began to rise. He had seen the captain of the *Brougham* being paid promptly, but no payment had been offered him for the use of his men and boats, nor for his own time over the past two months. He had his own buildings and affairs to see to, money he owed on his hotel in Wellington, he was worried about the non-appearance of the *Jewess*, and his precious horse had fallen into a Maori food pit and been killed. He had spent weeks putting himself out to fit in with the company men. Now Carrington was putting Barrett's rights aside for those of the company. His good-nature was stretched past breaking-point.

Yes, he openly told Carrington, he claimed all the land from this place to the beginning of the bush. No, he couldn't specify the acreage, but there was no doubt it belonged to his wife and his children. Carrington tried to demolish his argument by pointing out that his father-in-law had signed away the land when he put his mark on the Ngamotu Deed of Sale. Barrett retaliated. By tribal law, he said, his father-in-law could not 'convey the property from the children', quite contradicting his previous stand, that the company now owned all Ati Awa land in North Taranaki, and that those who had travelled away had forfeited all rights.

Barrett had not been really aware of all the implications, when he had arranged the original purchase, but suddenly, he was finding himself hoist with his own petard. As Wakefield's representative, he had happily 'bought' land for the New Zealand Company. Now, wanting to spread his wings on his own account in a fresh start, he realized that the land he had helped Wakefield buy, between latitudes 38 and 43 degrees, included his own. He had tried to persuade people in the Wellington pa to move off their land and let the Europeans have it; but now, Carrington was asking him to move off his land and let the Plymouth Company have it.[15]

Carrington expressed his outrage. Overworked and worrying about growing food shortages, he was frayed at the edges, too. And he had tripped over a tree being felled to create the bridge over the Huatoki River, had injured his back, and had been dragging himself around in pain for several days. While popular with his workers and unfailingly courteous to them, lavishing praise where he felt it deserved, he was sparing with the latter for Barrett, and at this stage displayed arrogance and prejudice towards him and his Maori relations.[16]

Barrett blew his top. In the only recorded example of his losing his temper, he came back strongly when Carrington challenged him. The surveyor noted it in his journal, complete with underlining to indicate Barrett's tone of voice: 'Well, Mr. C. possession is 9 points in law. The Coy would never have had a bit of land had it not been for me & unless they now reward me I am not going to give up this . . .'[17]

Carrington counter-attacked. The land Barrett claimed belonged to the Plymouth Company. By taking it, Barrett was 'making the Plymouth Coy reward him for services done to the London coy'. Barrett flung back about the 'reservation made for himself and Love's family'. How much was that, Carrington asked? Barrett couldn't (or wouldn't?) say; but intimated, incorrectly, that he could take 500 acres out of the middle of the town if he felt like it. Angrily, Carrington pointed out how bitter the farming settlers would be if they saw this, the finest land in the country, given over to Barrett; it would look like favouritism of the worst order. Moreover, why hadn't Barrett mentioned his holdings earlier, before he had let Carrington decide on this area for the settlement? That Barrett may have prevaricated was slowly dawning on Carrington. For Barrett, it was another unwanted pressure. After further exchanges, their anger vented, their tempers gradually simmered down, and the two men arranged to look at Barrett's claim the next fine day.

Barrett had reason to be on edge. Apart from the non-appearance of the *Jewess*, which was overdue with stores, the blankets and guns Carrington had requested and Barrett's whaling gear, there were suspicious goings-on in Wellington. The *Harriett* had been reported as undergoing 'a thorough repair', and was advertised in a February *Gazette* as available for passengers or freight going to the Bay of Islands, directing clients to '. . . apply to David Scott, Esq., or to Waters & Smith'. Barrett may have had pointed questions to ask his executor, David Scott, and his Will's witnesses, about expenses associated with the refit, considering his debts in Wellington. He needed to know those answers, as well as to supervise Hunt's management of the hotel, buy supplies, and put advertisements for his New Plymouth Hotel in the newspaper.

Accordingly, when the *Harriett* eventually called in at New Plymouth, on her way back from a trip to the Bay of Islands, Barrett took over as captain. He dropped off the Rev. Creed, who had been asked to intercede in the worrying Wanganui/Ngati Tuwharetoa conflict at Wanganui, and arrived in Wellington about the time that news of the *Jewess* was received. She had foundered at Paripari on 21 April, on her way to New Plymouth. Gone were the guns, blankets and Barrett's whaling gear, as well as much-needed provisions for the company and Jervis and Spencer's stores. And lives had been lost — Barrett's friend George Wade and Rawinia's middle-aged, rangatira 'uncle', Waiaweka, who had been returning at last to his ancestral lands. Their bodies were never found and the Waikanae iwi stripped the wreck as utu for the death of their chief. Moore, Wellington Carrington (Frederic and Ockey's brother) and others on board had survived. When Barrett returned home, his cannon fired a long barrage as a 'funeral dirge' for the dead, as well as, perhaps, for the lost cargo. Soon after, the guns were heard each morning, firing an 'on-deck' signal, after the survey team labourers protested they had no way of telling the time, when the foreman accused them of being late on the job.

Tempers were short all round now, as the period spent waiting for the allocation of New Plymouth sections was exacerbated by food shortages over several months. Carrington's last-minute purchases from the captain of the *Brougham* had been more than necessary, although the rat attacks on the stores had not helped. The seed potatoes being planted by the local Maori in ever-larger areas — anticipating 'many white people this year' — couldn't yet be lifted, however, and the wheat, turnip and other seedlings, still only green shoots above ground, would not be ready as food for months. The fowls and stock the settlers had brought were too valuable for breeding purposes to slaughter, and everyone turned their attentions elsewhere. The curlews and red bills on the beach, and pigeons and other forest birds became alternative casserole dishes to assuage the hunger pangs. Wild pigs were another alternative, but Ati Awa were after these too, and the barter price shot up.

Barrett's establishment also suffered. By May, for a month already he had been very short of salt and fresh meats and food in general for his whalers. He stopped their normal rum ration and requested other suppliers not to serve them. It may have been a precaution against the ill-effects of drinking on rather empty stomachs; but he could have been saving his funds. There was still no sign of the whales that were to be his source of income. As for his men, he did not have to pay them anything except their lay, but, as was custom, he had to give them credit in both his store and the company's. Unwilling teetotallers, his men tried unsuccessfully to bribe store keepers Jervis and Spencer to keep them supplied. Jervis had a secret supply of the 'coveted liquors' and one night, when he heard footsteps and someone trying his door, he thought it was the whalers. He called out, then panicked when no one answered and, alarmed for his stores, ran out into the darkness, sword in hand, calling for help. Soon the villagers and the Barretts and even the suspected whalers were aroused in confusion. Hands grabbed flares, and figures raced around in all directions shouting 'Waikato! Waikato! Taua!', till they realized it was a false alarm. It turned out that a visiting English 'gentleman' had 'thought it an opportunity for indulging in a little waggery', when he heard Jervis's 'Who goes there?' Barrett was not the only one who indulged in exaggerations and pranks. Nevertheless, for Jervis and Spencer, following the loss of their goods in the *Jewess*, it was the last straw. They packed up and headed back to Wellington, 'civilisation' and food on the table.[18]

The whales showed up in June, but the black mammals seen sporting quite close to the shore passed unscathed by Barrett's team. Cutfield reckoned that at least six or eight whales had been missed, and suggested a joint-stock whaling company be formed in New Plymouth, by-passing the Sydney merchants and profiting the community — but not necessarily Barrett. Cutfield envisaged the company keeping the profits from the oil exported, while lines, cordage and boats could all stimulate local employment by being manufactured in the town. He suggested an initial capital of £3,000, an amount adjusted upwards not long after, indicating the expenses Barrett was facing.

Quite unconnected with Cutfield's suggestion, Richard Brown arrived into this dismal New Plymouth on 27 June 1841. He had emigrated to the New Zealand Company settlement at Wellington and set up a store there, but later saw more potential in New Plymouth. He transferred his energies and established himself with a warehouse in the town and a whaling station at the mouth of the Huatoki, complete with four boats and a whaling-master from Queen Charlotte Sound. He was to be a solid competitor and neighbour of Barrett and, contrary to what some people may have thought, it appears to have been an amicable rivalry between the two men. At Te Awaiti, Love and Barrett, Jackson and Toms had helped one another, in the manner of the unwritten but binding laws of whaling, and Barrett and Brown were bound by the same ethics. When the settlers objected violently to the stench of Brown's first whale kill, he moved his whaling operation next-door to Barrett's. Brown soon had several whales to his credit when his next-door neighbours had none, an indication of Barrett's questionable management skills.

Barrett's whalers killed their first prey only three days after Brown landed, but had to anchor it because of rough weather and it floated out to sea and was lost before they could bring it in. Barrett, however, had reason to feel buoyant at the time. He and Cutfield had got together and agreed to have a 12-metre whaleboat built, to make the unloading of the next immigrant ships easier. Their mutual understanding was that Barrett should provide the timber, Cutfield the labour, and they would share and share alike.

On top of that, on Tuesday 13 July, after two months of stringent rations, two small vessels anchored in the roadstead, bringing food for the whalers in the form of sixteen or eighteen fine pigs and a quantity of potatoes, and a few stores for the settlers. They also left provisions and seed potatoes for next season; but when too few vessels followed in with winter supplies, those potatoes went into pots for dinners, putting paid to spring planting.

Good news from the outer world now percolated through that Her Majesty's Government had at last granted the New Zealand Company a Royal charter to settle New Zealand. Neither Barrett nor the settlers really understood the ramifications of this, but it was a boost for Barrett to have been part of the negotiations now being recognized by the Queen. What he may not have heard, was that the British Government had been informed of the increasing unrest between Pakeha and Maori, and was briefing a lawyer, William Spain, to go out to New Zealand as a Land Claims Commissioner to examine the dealings of this company it had favoured.

In July, the company erected a morale-boosting giant flagstaff on Mount Eliot that flew the Union Jack each day. Matching it, Barrett had a lookout pole set up on Moturoa, to enable his men to spot their prey earlier. By now, however, the whalers' period 'on the wagon' had ended, and they were drinking heavily again, with nothing else to do. At Te Awaiti, Bumby and Hobbs had found the men to be acceptable, compared with the 'licentious, drunken scum' at Cloudy Bay. Not so Barrett's team at Ngamotu. Old hands Bosworth, Bundy and Wright were included in Barrett's 'strange lot', as they were described in one settler's letter to his family in England.[19] The whalers became universally despised by the townspeople.

For Barrett, lack of a business partner, and Love's managerial skills in particular, may have been a critical factor. Barrett had flourished in the early days at Ngamotu, in partnership with Love, Te Puni and other Ati Awa chiefs. At Te Awaiti, he and Love built up equipment and workers, to become one of the major whaling centres. But Barrett on his own was dogged by problems. As a person, he was almost universally liked. His business acumen was suspect, however, when the failure of his freighting venture with the *Harriett* at Wellington was added to troubles over his hotel and inability of his teams to catch whales.

Not surprisingly, the whalers were all drunk on 10 July 1841, when the next chance of a kill hove into view in the passage between the Sugar Loaves. Barrett was reputed not to go out in the boats now, but when the mother whale and her calf appeared, he called up a group of Maori seamen to join him — and Richard Brown took an oar too — in one of Barrett's craft.

> On Friday Mr. Barrett went out in a boat with 1 white man and some natives (the rest of his whalers being intoxicated, a very usual circumstance) and struck a calf whale, this appeared to sober the men, as they manned the boats and went to Mr. B's assistance, they succeeded in capturing it and then went after the mother, which led them a long chase along the coast, they at last killed her and began towing her towards home, Mr. Cutfield went to their assistance with the Coy's boat, and they anchored the fish about 3 miles above this place; we saw them towing it along in the morning . . .[20]

By the time Barrett had harpooned the baby, his men on shore had seen this chance to earn their lay, and sobered up sufficiently to launch the second boat. They helped tow in the calf, then the chase was on for the mother, who, in the usual manner, had remained in the vicinity where her baby had been attacked. After a blistering 5-kilometre row up the coast, they harpooned and killed her, all the while being watched and cheered on by settlers covering every vantage point. Even Cutfield joined in with the long haul home, going out with a crew in the company's boat to give assistance. The prize was anchored for the night about 5 kilometres north of the Huatoki stream-mouth, then the next morning the whalers provided a spectacle for churchgoers at Creed's mission as they resumed towing the colossal black mammal back to the station. It was a boost for Barrett; but he would learn that whales caught off Taranaki going north — and having just calved — would be in poorer condition and give less oil than those he took at Te Awaiti. It would take much longer than he thought to pay off his debts. But whale steaks were very welcome for many that night.

Meanwhile, on top of a whaling season that had not started well and was not going to get better, bad news arrived for Barrett from Wellington. There were tangled complications in the lease for his hotel, suspicions increasing about the actions of some of his 'friends' there, and demands for payment of his debts. A grim time lay ahead.

*Chapter 12*

11 August–31 December 1841
DEBTS, DESTITUTION AND LAND DISPUTES

---

Barrett's bad news came just after the first anniversary of his daughter Mary's death, in the person of James Smith, of the Wellington store owners Waters & Smith. He had been acting on Barrett's behalf since he had left Wellington, with the hotel as surety. Smith arrived in the little 27-tonne schooner *Surprise*, which he had chartered, filled with flour and rice and sailed north, calling in at Wanganui to sell these provisions en route. He did not bring them to Barrett at New Plymouth. Smith's business with Barrett was about settling overdue accounts, not adding to them. Evidently, David Scott had somehow let Barrett down, and Smith had had to pay out a considerable sum to cover the debts that had mounted up in Barrett's name.[1] Now Smith wanted Barrett's Hotel as recompense. In what was almost adding insult to injury, Smith decided to open a branch store in New Plymouth, after his two days there.

Barrett had little time or advice as to how to react. He sent a letter to Wakefield when the *Surprise* sailed again:

Moturoa, 14 August 1841
Dear Sir
Having been greatly deceived by Mr David Scott by false Promises of support and who has brought me deeply in debt to Messrs Watters & Smith — who have Paid debts in Port Nicholson for me to the Amount of nearly 800 pounds and given me goods for 500 pounds more so I have been obliged to sell Messrs Watters & Smith the Hotel for 1,500 pounds in consequence of my poor season not being able to pay them in Oil as I expected having only caught three Whales — along with many other Misfortunes

## Debts, Destitution and Land Disputes

I have has a friend of mine to request you will give Messrs Watters & Smith that protection and not allow any one to disturb them the Hotel being built on a Native reserve — I have been reccomending them to put my Old friend - - - Blackman in charge of the Hotel but Mr Smith says that could not be done without your consent.
I remain Dear Sir
Your well wishing friend
Richard Barrett[2]

*The only known authentic letter written by Barrett, now missing from National Archives. Two others held there are in different handwriting, and their signatures are different from this one, and those appearing on the seven pages of Barrett's will.* NATIONAL ARCHIVES

Of the misfortunes that had plagued Barrett, one was probably the loss of his other whaleboats (usually pulled up on the beach in front of his Wellington house when not in use), perhaps seized by Waters & Smith in compensation, or destroyed in the wreck of the *Jewess*, along with a lot of his whaling equipment. Another may have been the refusal of Carrington and Cutfield to pay any dues for the use of his houses, men and boats; and there may have been emerging resistance from the villagers, despite his wife's connections, to his intended use of extensive acres of land around their pa. With all these Pakeha around him, he was not the Tiki Parete they remembered from the old days.

Five days after writing, Barrett followed his letter to Wellington. By now, the town of tents and reed whare he had left seven months earlier had about 3,000 inhabitants and many more permanent houses. The waterfront was lined with buildings extending onto the so-called Lambton and Thorndon 'flats' behind, a new wharf was taking shape in front of his hotel, and new streets had appeared, with names like Molesworth, Willis, Buckle, etc., all connected with New Zealand Company directors. A road now linked Thorndon with the Hutt, and just north of the town, John Wade and his business partner, James Watt, had constructed a dray road to service Wadestown — the one- and two-acre sections they had opened up on the slopes for the continuing flow of new arrivals. Domestic animals, crops and fruit trees were everywhere and Wellington was no longer as dependent on Maori trade as before.

Other changes were not so positive. Unemployment, poverty and alcoholism were evident in a large number of immigrants, as they waited in varying states of anxiety for their land to be awarded to them 'legally'. The influence of Te Puni and Te Wharepouri was waning, and Te Ati Awa at Pipitea, the two hapu of Taranaki iwi at Te Aro and Ngati Toa at Porirua were being completely unco-operative. Some settlers were becoming aware of possible distasteful aspects of the New Zealand Company's negotiations. Others were simply wondering what had happened to those genial, welcoming villagers they had first encountered two years previously, who now demanded the same rates of pay as

Europeans and were selective about what they did. The Maori, for their part, having endured the bulldozing and increasingly threatening tactics of the Europeans for eighteen months, were still loyal to the Queen in England, but did not care for many of her subjects.

On the same day Barrett arrived in Wellington, Governor Hobson sailed in on the government brig *Victoria*, on his first official visit. He brought a varied group with him: Felton Mathew, the acting Surveyor General, Edmund Halswell, who was to be a Senior Magistrate, George Clarke snr, Chief Protector of the Aborigines (who was to be Halswell's *bête-noire*), Edward Shortland, Hobson's newly appointed private secretary and brother of Willoughby Shortland (who was still Wellingtonians' *bête-noire*), and Te Wherowhero.

Hobson showed the effects of his stroke in his movement and speech, but the New Zealand Company men made no concessions for his frailty. Some of them referred to him as 'our imbecile Governor', while others could not stand the idea of his appointment, and had already sent a memorandum to Her Majesty's Government seeking his recall. Wakefield, equally as vituperative as his fellow colonists behind the Governor's back, met Hobson graciously, and accompanied him watchfully.

The Governor disregarded the animosity he sensed among the settlers and concentrated on the complaints of the tribespeople. He listened to Te Wherowhero and, in a judgement which may have riled Te Ati Awa, agreed that 'Ngati Awa [*sic*] are slaves and cannot sell land', that, therefore, Te Wherowhero was the 'sole owner of Taranaki', and paid him £250 to drop that title. Waikato could therefore no longer claim to have rights in New Plymouth.

Then Hobson, Wakefield and Clarke paid a formal visit to Te Aro, now disfigured by a large European wharf in the middle of the pa, and listened to the chiefs' grievances. The colonel had accompanied Hobson specifically to counter these. He dismissed their grumbles, telling Hobson that they had been lucky to be paid anything for the land they occupied, since they were mere slaves; but the chiefs irately refuted Wakefield's words, and Clarke supported them strongly.

After this, it came as something of a surprise that Hobson's eventual findings favoured the New Zealand Company. This captain, who had impressed his naval superiors, soothed the northern Maori tribes and 'lawless' British citizens, side-stepped the threat of French colonisation and overcome the debilitating effects of a stroke, was no match for several days of Wakefield's matchless charm and doublespeak. When Wakefield conceded that the first 'New Zealand central transaction', of two million acres (see Chapter 7), was not realistic, and abandoned any claim to it, Hobson issued a schedule affirming a generous part of the rest of the purchase that Barrett had engineered two years before. What was more, Hobson left Wakefield with a confidential note, to the effect that the government would go along with any strategy Wakefield employed to get possession of the lands mentioned in the schedule, short of using force!

Barrett, on the other hand, did not have anything like the same success in Wellington as Wakefield. He discovered that everything Smith had told him about his finances was inescapably true, and that any attempts to discharge his debt were distinctly complicated by the fact that Barrett's Hotel sat on a Native Reserve. He left his affairs in lawyer Hanson's hands and returned to New Plymouth after only a week away. He was not aware for a while, therefore, that just after he left Wellington, it was decided that the next New Zealand Company settlement, of Nelson, should be formed at Whakatu, or Blind Bay, the harbour from which Barrett had so carefully steered Carrington away.

By now, the settlers in New Plymouth were moaning vociferously about having ever left England. Living was a continual battle to shield themselves against the elements and their food supplies against termites, insects and hungry animals. Drunkenness was rife among the labourers, in a dreary existence with too little to do. Flour supplies had run out again and there was no likelihood of more until the next boatload of settlers arrived. Te Ati Awa, too, were hungry. The co-operative ones had planted more crops than usual, to feed the coming Pakeha, but so many more Europeans had turned up than they expected, that they also were short of food.

On Friday 3 September 1841, Captain Henry King, the official Agent for the Plymouth Company, landed and took over from Cutfield. He was fifty-eight years old, and, with his son, had rowed 32 kilometres from where the next immigrant ship, the *Amelia Thompson* was hesitating, waiting for the right weather to come in and anchor. It was the hungry settlers' first intimation that supplies were almost within reach. However, its Captain Dawson had heard stories about the dangerous roadstead at New Plymouth, promulgated, Carrington guessed, by traders who knew they were onto a good thing in the early 1830s and spread stories about the coast being 'most awful'.[3] Dawson was taking no chances over the safety of his passengers and cargo.

No chances indeed. For the next seven weeks the little community waited on starvation rations, while the ship, with fervently awaited white flour in her hold and billowing white sails above, did a maritime minuet, advancing, retreating, then advancing again, sending off a boat or two, then mincing off again, beyond the horizon, repeating the performance again and again, but never committing herself to anchor.

In his constant but unofficial role as harbourmaster-cum-pilot, Barrett had gone out to the ship on Sunday 5 September. He had assured Captain Dawson that the weather was due to be settled for a week, and it would be safe to disembark his passengers. They had been on board for more than five months. Barrett brought back some of them in his craft, towing more behind in the ship's longboat; but Dawson procrastinated 'faintheartedly'.[4] He sent in just one boatload of immigrants that day, from south of the Sugar Loaves. They swept up the beach in a perfect Sabbath landing, after five hours of rowing and sailing, to the warmest welcome possible from the Maori, and from the Pakeha, including Barrett, all aching for news of home. Another boatload was landed on Monday the 6th, but, Carrington noted later:

On the 8th there was very little wind but very heavy rain all day till 5 o/c p.m. at which time I saw Mr. Barrett. He told me the *Amelia* was laying 'much in her old place', 8 or 9 miles the other side of the Sugar Loaves — he tells me he has named it to Mr. Cutfield but he has no orders to bring her to the anchorage & unless he has orders he shall not again go off to the vessel. I was sorry to hear this & told him . . . how I should act if I had the power. He told me he would have half the things out of her if he had been allowed to pilot her. I am convinced he would.[5]

Even so, Barrett kept an alert eye on the ship, and swung into action when he saw boats leave her at 5.30 p.m. on the Thursday, 9 September. They were at least 13 kilometres out and it would be dark before they reached land. He had men light a beacon fire or two on the beach, between the two reefs at either end, for the boats to aim at as a landing spot; but the tide was low and the boats, coming in tandem, struck the rocks 'some distance out' and were 'stove'. The Maori and whalers watching on the beach seized firebrands and waded out into the fortunately calm water. They brought ashore the drenched women and children who constituted the party, their '. . . bedding and belongings thoroughly saturated . . . money and goods . . . lost', but no lives. From the beach, they were taken to 'the warm welcome of Barrett's establishment', which seems to have been one of the large shelter houses.[6]

In the days that followed, passengers continued to come ashore in dribs and drabs, most of them being initially accommodated at Barrett's, which was as close as he ever got to running a New Plymouth hotel. 'While stopping at Barrett's we sometimes had fresh pork or fish, but the stand-by was salt pork, generally with a whaley flavour. . . . Potatoes and Maori cabbage were the vegetables. The latter was very good. It was of the Swede family, and had an equally good taper turnip root.'[7]

It would seem that these immigrants heard about Barrett as a character and practical joker as they became part of the local scene. On 17 September, a week after the unfortunate wet landing, an earthquake shook the community. For those on their feet, the ground shook like 'the interior of a flour mill' for a full minute; for those who had retired, it was like having 'two or three stout fellows . . . under the bed alternately pushing up the sacking with their backs ...'. People were heard to wonder whether it was this Dicky Barrett fellow, or a wag, or a drunken reveller, '... giving the house a good shaking from the outside'.[8]

The *Amelia Thompson* did not complete her unloading until six weeks after her arrival, but at last, on 30 September, all the baggage was ashore, and the 150 barrels of flour were beached for the impatient settlers. These, together with the salted meats and flour which came shortly afterwards in the company supply ship *Regina*, put paid to hunger pangs in the interim.

Barrett may have had to share the blame for the delay. Going right back to the late 1820s he, and other flax traders on the west side of Te Ika a Maui, had supplied disturbing descriptions of the exposed coastline, in an effort to keep competitors out. The trading ships too, which serviced them, had connived in this deception for their own benefit, and rumours now reached all new seafaring captains to treat the area with circumspection.

Other rumours about the destitution at New Plymouth were reaching the ears of intending settlers and many potential immigrants, particularly the prospective employers, disembarked instead at Port Nicholson. As a result, while Wellington had unemployment, because the future employers had no land title that would encourage them to engender jobs, New Plymouth simply did not have enough employers, and Cutfield found himself with hundreds of labourers, guaranteed jobs when they left England, on his company books.

Whatever their class, the influx of new settlers and their luggage increased Maori antagonism. Instead of fearing a Waikato invasion, and having to put up with the Europeans as an encumbrance necessary for defence purposes, the local villagers now began actively to resent the continually increasing numbers of settlers. The resultant tensions must have been unpleasant for Dicky and Rawinia and for other couples who had intermarried.

The well-balanced food chain that Te Ati Awa had known previously at Ngamotu had also been disrupted by the influx. Food shortages became endemic. Pioneer life was onerous enough without lack of calories as well, and, by the beginning of October, the leaders of the community were bickering. They too had been affected by the short rations, were coping with demanding events, with interruptions to the leadership role, and with feeling abandoned. Carrington was incensed over his workers' poor performances and his own working and housing conditions; Cutfield had been demoted by King, and King found himself rather out of his depth. Carrington and Cutfield were clashing, and both found King unsatisfactory.

Things came to a head one day when all three clubbed together against Barrett. On Saturday 2 October, Cutfield and King asked Barrett to come and discuss with Carrington the recently completed 12-metre whaleboat which Cutfield and Barrett had agreed to build together. Cutfield and King had their heads together when Barrett and Carrington arrived at the company's storehouse for this consultation, and it became apparent they were aiming to squeeze Barrett out of his share of the vessel. The two men jointly asserted it had been built for the express and sole purpose of discharging company cargo. Barrett indignantly repudiated this. They persisted with their claim. Barrett challenged Cutfield resolutely, and the latter finally caved in, in front of Carrington, admitting that half the boat did indeed belong to Barrett. The surveyor, himself, then challenged Barrett about the accounts he had rendered for the use of his 'shed', stating that he considered them outrageous. Opportunely for Barrett, a man whose opinion the surveyor valued was nearby, and he declared that the figure Barrett had mentioned was 'a fair colonial charge'. Carrington grudgingly agreed to pay it.

Barrett then retaliated. While he was carrying a load of debt, he felt he had considerable debts *owing* to him. Evidently, at the time the *Brougham* was being unloaded all those months ago, he had been unwise enough to waive an offer of payment. Now he was having second thoughts. He demanded reimbursement for the use of his men and boats from Carrington; but Carrington refused. He knew that the Plymouth Company was facing difficulties at this stage (although the general public were not informed) and he could not take the responsibility for lumbering it with this delayed charge. He dismissed Barrett's claim, saying that, anyway, he had given Barrett's men £3 and some grog then, although they had done very little. Barrett protested that he had to pay the men monthly wages and maintain the boats, and was losing money. Carrington heaped coals of fire on Barrett's head by offering to pay the cost out of his own pocket, whereupon Barrett offered to withdraw his claims. After further unsatisfactory dialogue between the two men, both coping with severe financial stress, the 'charge was allowed to stand good'.[9]

Only three days after these astringent exchanges, one of Barrett's antagonists was replaced. The New Zealand Company had absorbed the financially exhausted Plymouth Company earlier in 1841 and sent out their own representative, Captain Liardet RN. He arrived in New Plymouth on 5 October, on board the *Regina*, so King now became redundant.

Liardet may not have been impressed by one contact he had with Barrett's establishment during his time. Wakefield had recommended Barrett to Liardet, and when Liardet saw a ship, caught in a sudden gale, tossing dangerously at her moorings, he rushed to Moturoa, looking for seamen to help; but, to his 'disgust', the whalers were 'panic struck by the ... weather ...' and he could not get a boat crew to go with him. It seems that eight years with copious supplies of arrack rum had diluted the old hands' resolve and strength, and that their example affected the newcomers. In that whaling season, such as it was, Barrett's team took only three whales, and results did not improve much over the next five years, for all Barrett's good humour and general popularity, demonstrable quick-thinking and courage. The whalers' performance under his leadership was repeatedly substandard, compared with when Love was alive. Liardet, on the contrary, exhibited impressive competence, but New Plymouth's misfortunes continued when a cannon exploded in his face only six weeks later causing serious injury and the townspeople were forced to rely on their earlier, less-effective leaders.

Barrett probably hoped for profits from farming to back up his whaling, but it would take two or three seasons and much labour to achieve a return. There was only primitive equipment for clearing the land, establishing good pasture, fencing, ploughing and planting, and it was all complicated by ever-present sandflies, mosquitoes, caterpillars and grasshoppers, as well as threats to stock from disease, accidents and tutu poisoning, and bad weather. What with the failure of the whaling season, the whole affair seemed guaranteed to break not only his back but his heart; but he is never described as downhearted. Perhaps Rawinia and the girls' influence helped. They were already producing an abundance of vegetables in their garden, making the Barretts household almost self-supporting.

His good humour must have been stretched often, however. At times his shelves were bare like everyone else's, but he had more opportunity (if less credit available) than most to replenish his stocks. Although people at New Plymouth felt completely cut off from either land or sea access to Port Nicholson, there were little coastal vessels coming and going with moderate frequency. The captains of these ships were familiar with the habits of the sea on that part of the coast, and supplied small services and carried provisions from port to port. Barrett not infrequently went in them to Wellington.

One of these trips was in late November, just after Carrington had finally completed his town plan and the settlers at last had been able to choose their town sections. With all rations in short supply again, Barrett went to Wellington to reprovision, also hoping to salvage something from the ruins of his Wellington venture. Hanson, to whom Barrett had turned originally, was now representing both Waters and Smith. So Halswell, the English lawyer who had come to Wellington with Hobson and been subsequently appointed Protector of Native Reserves, and to whom Hanson had originally written about Barrett's predicament, took over as Dicky's legal representative.

Barrett was resigned to losing his hotel to Waters & Smith, but he asked Halswell to try to retain his little house and the fenced land on which it stood.[10] Through Colonial Secretary Shortland, Halswell wrote on Barrett's behalf to Hobson, now Governor in his own right, pointing out he had been unable so far to lease this Native Reserve, and requested that Barrett should retain the home and property 'in right of his wife a native woman of some consequence by whom he has a family'. Unwisely, he quoted Wakefield's support for this application.[11] The Governor did not approve of the way Wakefield had meddled in government affairs and, through his Secretary, wrote back astringently that he:

> ... distinctly disputes the right of the Agent of the Company to dispose of any reserves and in the event of any opposition being made to His Excellency's disposing of the land in the way he may judge most equitable ... will claim the Allotment for the Natives with all the buildings upon it.[12]

In his reply, Hobson laid down arbitrary terms for a seven-year lease for Barrett, who was required to pay 5/- per foot for the land on which the hotel stood, and 20/- per foot for some adjoining ground. This all meant that Barrett had to contribute £54.5.0 per annum to the Crown, to lease his own hotel, so that he could issue a sublease to Waters & Smith. An unexpected blow was a demand that Barrett also pay a year's arrears. It was not an attractive solution for Waters & Smith, but it was better than a complete loss on the monies Barrett owed them. They took over the lease and the hotel and ex-manager Hunt took his family to Valparaiso. Barrett found himself saddled with even more debt, with still no prospect of repaying it until he caught more whales.

Waters & Smith as a partnership eventually also collapsed financially. One reason for this was that when the market in Wellington, and Waters & Smith's store, became 'glutted' with goods, Waters could not send them to his partner at New Plymouth, since none of the ships in Wellington dared sail into that supposedly dangerous anchorage.

Gradually, however, the New Plymouth roadstead did prove its worth as a feasible port of call, as Barrett had promised. There were difficulties with wind and waves, and 1841 saw frequent delays because of an unheard of number of nor'westerly gales and heavy surf; but wind changes were usually foreseeable and, by the end of the first year, there had been no loss of life due to the sea, and only one ship, the *Regina*, had ended up on the beach.

The situation regarding the town was not so positive. Carrington's eventual plan covered 550 acres and, instead of being in a tightly knit community, the citizens on their newly chosen town allotments found themselves sparsely sprinkled on those acres among vacant sections belonging to absentee owner speculators. The only time they came together was on Sundays, when the townspeople walked to church at Ngamotu, and its tiny community of mission, whaling stations and kainga, isolated during the week, was inundated by the 100 or 150 adults who attended Creed's chapel and the sixty or so children at Mrs Creed's Sunday School.

Dicky seems to have been a God-fearing man, like most of his generation. He certainly was a man of humour, and, as well as becoming a legend with his practical jokes, he created legends with his stories. For the benefit of a colonist writing a book, *Sea and Sky, or a Trip to New Zealand*, Barrett described how a hefty grindstone was brought to shore.

> It belongs to my father-in-law. He took a fancy to it on board a Sydney schooner which came to trade with the tribe. The Captain in a joking manner said to the natives 'you can have it if you will roll it to the shore'. The schooner was lying about a mile and a half out, but before the Captain could prevent them the natives had lifted the grind-stone and let it drop into the sea. The chief and his men then dived into the water, and one after the other in turns rolled the grind-stone to the beach. The Captain lost the grind-stone, but he had its worth in the fun he witnessed as the natives one after the other, came bobbing up to the surface of the water panting and blowing after their exertion in trying to move the stone to the beach.[13]

That stone was an integral part of one of the first mills to be built, as the settlers began at last to construct their permanent homes and businesses. The surveyors, having completed one job, now began preparing the country sections, which were to be spread out across a wide band of territory and into the Waitara district. Toponunga[14] and a fellow chief from that area came to confront Carrington on 15 December 1841. They had returned from Waikanae to deny absolutely the sale of the Waitara land. There had been some payment received, they had heard, but it was only for 'some scattered lands between the Waiongona [*sic*] River and the Sugar-Loaves,' they said, and definitely for none of theirs.

The next day, Carrington took them to confront Barrett, who had to spend from 10 a.m. until 3 p.m. defending his past performances in front of the two chiefs, as well as Carrington, Creed and Cutfield. Barrett stuck to his guns about his purchase and said that Toponunga had not been there: he had been a slave.

Toponunga was furious. He had not been a slave; he had been at Waikanae when Wakefield called in, in 1839, and was still there when Barrett was negotiating for land at Ngamotu. Creed spoke up heatedly 'in the cause of the natives', Barrett defended himself, and they all argued round and round in circles. Eighteen months later, Barrett admitted his mistake to Carrington. Initially, he said, he had reckoned that all the land mentioned in the Ngamotu Deed had been fairly purchased; but then he began to realize that too few owners had signed and that Toponunga had been in the right.

Another threat to his purchase seemed imminent when it was heard on 25 December that a band of about 100 Waikato was approaching. There was general consternation, but on arrival Waikato announced they had come to make final peace, and that was how they acted. After speeches and food in town, they moved down to Ngamotu and attended the Boxing Day church service in Creed's chapel. Three more days of feasting and long hours in koreorero with their friends and relations at Ngamotu followed, before they performed 'an amusing specimen or two of the war-dance'[13] in front of Richard Brown's storehouse, and headed off north again on 29 December. The spate of words in 1841, and Hobson's payment to him, had apparently achieved what Te Wherowhero may have been seeking in 1833 at Otaka pa — a peaceful solution.

*Chapter 13*

13 JANUARY 1842–27 AUGUST 1843
NO LONGER CENTRE-STAGE AND
THE COMPANY FALTERS

---

The summer produced no tangible progress for Barrett. He was still agent for *The New Zealand Gazette & Wellington Spectator* in New Plymouth, but he was noticeably and increasingly excluded from town affairs. In the first instance, he was passed over when the official position of harbour-master and pilot was created, despite all his voluntary contributions. This may have been because he lived too far away from the centre of town; but that had not prevented his services being used previously. In another omission, he was left out when the townspeople organized a petition at the beginning of 1842, asking the directors of the New Zealand Company to form a safe anchorage, and quoting Barrett as 'a resident whaler here' who advocated the necessity for improving the facilities. Two sets of anchors, chains and buoys arrived, which, once in place,[1] improved the safety of the roadstead; but, here again, Barrett lost out. The moorings were positioned so that larger vessels landed their goods and passengers in the small natural basin at the mouth of the Huatoki, instead of at Barrett's front door. The New Plymouth community, which had its own problems with alcoholism, preferred that new arrivals were not introduced to the town via Barrett's malodorous whaling station and its disreputable whalers.

The health of Barrett's comparatively well-disciplined whaling team from Te Awaiti had deteriorated dramatically and, by the end of that summer one of them had already died from delirium tremens. Despite being spurned, Barrett, with his typical generosity, still made his cannon available to answer any salutes fired by visiting ships; but this was a very minor contribution to company affairs, compared with his previous involvement since his first meeting with William Wakefield.

Richard Brown next door, who had helped to compose the letter to the directors and was quite well educated, had advanced quickly in the community. In addition to running a well-patronized store and catching whales, he was active in civic affairs and was elected to town committees. By contrast, Barrett, after being one of the kingpins of the community's first few months, languished out on a limb at Ngamotu; but he and Brown remained friendly.

Dicky may have sown the seeds for his decline in popularity by his earlier stratagems. It was becoming known that 'a certain person' had prevented Carrington from examining Nelson Haven, with its safe harbour, as a possible site for New Plymouth, leaving the surveyor with Taranaki as the only choice. The community was still on reduced rations and enduring other hardships, and the depression these deficiencies produced, plus the gossip about Barrett and Nelson Haven, may have been enough to sour people generally against the whaler. Even so, about this time, the townspeople adopted a rousing, patriotic song, composed by one of their number, one verse of which mentioned Barrett as one of the features of their New Plymouth:

> ...To strike the whale with harpoon true,
> We've Barrett and his hardy crew;
> Our flagging spirits soon we cheer,
> With Secombe's stout, and George's beer;
> Nor fetch tobaccos from afar,
> Whilst Nairn can twist the mild cigar ...

One feature not mentioned in their song was an unexpected plague of rats that descended on the town in February. People were accustomed to seeing dozens of them in a day, but now seething thousands darted around, invading house, tent, whare and store. Voracious as the locusts of the year before, they ate everything they fancied, and demolished an entire crop of 800 melons that Rawinia and her daughters had grown. Surprisingly, they left standing the large block of healthy maize (Indian corn) that Barrett had planted. They provided an astonishing phenomenon to greet the next

200 or so Plymouth Company settlers who arrived late that month in the *Timandra*. In just over two months, one of these killed about 500 rodents in his tent alone, before they dwindled and were gone, leaving myriad paw marks on the beaches, all pointing in the same direction as they migrated south, as though the Pied Piper had tootled them away.

In April 1842, just as the long-awaited allocation of New Plymouth suburban sections began, Wakefield came north in the *Brougham* on an official and conciliatory visit, to see how Carrington and the town's inhabitants were faring. It was the first time the Principal Agent had actually stepped on New Plymouth soil. Carrington met him courteously, with up-to-date survey details; but the householders met him with bitter remonstrance, moaning that, like Hobson's coming to Wellington, he was a year late, and they all felt cut off, as if they were doing a life sentence in a penal colony. After four uncomfortable days, Wakefield left, promising that funds would be made available to cut a bridle path through the bush to open up communication with Wellington.

Barrett did not need this land route to get away from the isolation and, within the month, he was back in Wellington, picking up the usual supplies and, still determined and optimistic, searching for whaling tackle; but this time he also signed a codicil to his will, on 25 April 1842, replacing Scott and Brewer with John Wade and William Guyton as executors. The story behind this decision is lost, although there were previous indications of friction with Scott over the pulling down of his house by Barrett and Wade (see page 166). Wakefield may have brought other, more recent reasons when he visited, which may have led to Barrett changing his executors.

Barrett had left Wellington before Commissioner Spain's court opened there on 15 May. The hearings, to investigate the legitimacy of Wakefield's and Barrett's purchases for the New Zealand company, took place in the Courts of Justice, in an atmosphere thick with European unease over increasing Maori opposition to land sales, and Maori suspicions about the European invasion. At the beginning of the proceedings, 200 to 300 tribespeople and several dozen settlers sat in the courtroom

*William Mein Smith painted three masterly landscapes of Wellington in about 1843. This aspect looks to the southern side of Lambton Harbour.* ALEXANDER TURNBULL LIBRARY

each day. Wakefield and the rest of the Europeans expected Spain to issue a pro forma judgement in favour of the New Zealand Company. The Maori hoped for just the opposite. Right from the start it was plain that the going would not be straightforward for Wakefield, as the Maori chiefs spoke out strongly for their own case and made no bones about what they thought of the colonel.

Wi Tako was an early witness, and denied selling land. In one burst of words, he challenged Wakefield:

> I ask you Pakeha what did the Queen tell you? Did she say to you 'go to New Zealand and fraudulently take away the land of the natives'? You say no, then why do you encroach upon land that has not been fairly purchased?[2]

Taringakuri, from Kaiwharawhara pa, not heard from since he invested Otaka pa with his warriors, put in his spoke. He corroborated some of Wi Tako's evidence and added morsels of his own: that the colonel had bought the land only from Ngauranga to Pito-one, that the payment was for the sea, that the Europeans had been unpleasantly dismissive and said, 'Go and live in the mountains; live in the sea', and that the Taranaki people at Te Aro got nothing because Wakefield was stingy. Wairarapa, from Pipitea pa, denied any intent to sell. Te Puni was evidently the most supportive of the New Zealand Company's case, although at times even his evidence seemed evasive or contradictory.[3]

After a time, Wakefield retired 'in a paddy'. He was irate that Spain was giving the Maori position so much prominence, and said he was 'too tired to continue'. More and more tribespeople filled the Court, but the Europeans, as the weeks passed and they discovered that Spain was actually running his Court 'as though it was in England', denigrated its purpose and drifted away.[4]

Sporadic reports of all this reached New Plymouth where, in early May, John Wicksteed was sent from Wellington, with his wife and family, to take injured Liardet's place. Wicksteed was a 'clever, energetic sort of fellow',[5] or 'a vulgar hound',[6] depending on who was describing him. A fluent speaker and writer, he had been connected with the London *Spectator* newspaper in earlier days, but had come to New Zealand, arriving on the *London* with Carrington, as agent for the Church Society to administer land set aside for the establishment of a bishopric in New Zealand. He moved up from Wellington to Taranaki and took over control as a thrifty Englishman, dedicated to the Empire, aware of the financial problems the company was facing and apparently full of confidence in his leadership abilities. Within a month, he had pushed the community of mixed races to the brink of war.

He started on a positive note, sending two of Nairn's sons to engage gangs of Maori to begin cutting two halves of Wakefield's promised track — one beginning at Patea, the other at New Plymouth, to meet halfway. Unfortunately, his following actions were not so productive.

*Part of Ngauranga pa, showing a monument to Te Wharepouri, who was buried at Petone.* ALEXANDER TURNBULL LIBRARY

Carrington, who had chosen to go no further than Waitara when laying out the farming hectares for the settlers, had become more empathetic towards Te Ati Awa. Now aware of how highly they valued the Waitara district, he had promised them two favourite spots there as native reserves.

On 20 June, however, when the selections for rural lands took place, no such reserves were specified. Wicksteed ignored the subsequent protests, completely disregarded Carrington's promise and allowed the settlers, one of whom was the magistrate, J. G. Cooke, to choose any prime sections they liked on either side of the Waitara River. Wicksteed contended that the tribespeople would far rather live together somewhere else, than separately in the two different areas Carrington had designed. Te Ati Awa in that region made their preferences clear. Previously, they had indicated they did not wish to live close to the whites; now they made it obvious that they did not even want the whites to live in their district. When they found Europeans living, and surveyors stomping, on their prime lands, they began to drive off the latter and attack the former.

*John Wicksteed, New Zealand Company Agent in New Plymouth.* ALEXANDER TURNBULL LIBRARY

Wicksteed continued to exacerbate the situation. He agitated and dissembled, saying that Te Ati Awa were slaves who had no claim whatever to the land. He saw the use of force against 'natives' as quite natural, and swore in a number of armed constables. Then he put them under the command of the hardly impartial Cooke, who transported them all to Waitara by boat, fired volleys over the heads of the protesters and 'in the name of the Queen' took possession of the land Carrington had promised them.[7] Cooke, and many in the community, had persuaded themselves that Te Ati Awa who had moved away had lost their land rights in Taranaki. They would surely never have admitted to losing their own land rights in Great Britain through becoming colonists.

The startled chiefs at Waitara came back at Cooke. They repudiated absolutely the deed that Tiki had negotiated on Wakefield's behalf. They were Puketapu hapu of Te Ati Awa, totally separate from Rawinia's Ngati Te Whiti, they stressed. If any land had been sold, they emphasized, it was only in the Ngamotu area. Cooke ignored their claims, rebuked them, and talked sternly to them about arrests and trials; and

Wicksteed seemed pleased with himself when the chiefs, though full of bitter resentment, arranged to wait and see what Spain decided. Wicksteed pontificated afterwards that there would be room in the reserves for twenty times the present number of Maori, so why were they worried?

Many Ati Awa hapu around New Plymouth were drawn into these confrontations, which brought Barrett into opprobrium from both sides, he having sworn that he had bought the whole district. From this time on, as well as being partially ostracized by his fellow countrymen, Barrett steadily lost the mana Te Ati Awa had accorded him as a European during his first decade in New Zealand. They now treated him as 'one of themselves', which meant he was not omitted from everyday disagreements and unpleasantness. At the same time, when it suited them, they identified Barrett as a white man, and included him in Pakeha unpopularity when the settlers acted unwisely.

These events occurred during what should have been Barrett's whaling season. It had opened in May/June, but Barrett's men caught nothing until 7 August, and then only one whale. His acquaintances in Wellington would have been surprised to hear this, since a group of visiting Ati Awa had let it be known that Tiki had taken eight whales and that the living was easy; but they had confused Barrett with Richard Brown, who contributed most to the meagre twenty-two casks of whale oil (worth only £88.4.6) exported from New Plymouth that year. A season with one whale was not viable, and Barrett actually shut down his whaling operations at that point. However, a Richard Rundle, who had arrived on the *Amelia Thompson* with his wife and about seven children, had money to invest in his new place of residence. He chose to put some of it into a flour mill with a partner, and some into local fishing and 'his financial backing enabled Barrett to resume whaling.'[8]

Whether the Barrett farm produced profits at this time is open to conjecture. Barrett was only just beginning to establish his crops, and would have been going through difficulties in farming as well as whaling. Along with the general community's crops, Barrett's clover and swedes were tempting fodder for the grasshoppers, his wheat for the worms, and he and his family for the mosquitoes. All the farming people complained. Working the land never was easy.

But more cattle were coming into the district, allowing diversification and, hopefully, better returns in the future. Despite the difficulties caused by the lack of roading from Wellington, Cooke, Richard Brown and various other entrepreneurs, were importing animals, and the Hongihongi lagoon on Barrett's acres was a convenient watering hole for any animals he stocked. Fences on his property, to contain any beasts he could afford to buy, would have to be added to his farming agenda.

Over these months, each time Barrett went to Wellington, his old friend Te Wharepouri was becoming progressively debilitated. Irascible and uncertain since having lost control of his land, he was now also losing control of his European settlers, who were deserting his area of influence at Pito-one and Ngauranga and moving to Thorndon,[9] and he sensed that the Governor saw him as a leader who had dealt in land which was not his. His mana receded, his imposing presence crumpled and he turned to alcohol as a prop. On his downhill path, 'that restless fighting devil'[10] stirred up unnecessary trouble between Te Ati Awa and the Wellingtonians in 1841. The following year, Te Wharepouri became very ill with an abscess in his ear and died on 22 November, probably unaware that Hobson had predeceased him by about eight weeks. Both men, still in their forties, were possible victims of colonisation stresses. Te Wharepouri's body was buried at Pito-one, with part of his magnificent canoe erected at Ngauranga as a memorial, like Love's at Te Awaiti.

By now Wakefield was also losing his way in some quarters. Described in 1840 as this 'most kind-hearted man ... gives universal satisfaction ... greatly beloved by the natives',[11] by 1842 colleagues were fulminating against him. Not able to penetrate the iron reserve he maintained beneath his superficial charm, they were also irritated and alienated by his apparent lethargy. '... he has only just energy to sit down and write an excuse for not having energy ...',[12] one of his associates wrote, while George Clarke jnr described him to his father about now as a designing, crafty, unprincipled fellow.[13]

*The town of New Plymouth, as sketched in 1843 by Emma Wicksteed, wife of John Wicksteed, the New Zealand Company's Agent.* TARANAKI MUSEUM

Barrett's life, like Te Wharepouri's and Wakefield's, continued under gathering clouds. Little good fortune came his way in 1843, even though the new settlers arriving on the *Blenheim* and *Essex* in late 1842 had increased the number of Europeans in New Plymouth to 900 — more than double the Ati Awa population of 400. Despite the town proving to have a safe roadstead for the increasing number of little coastal vessels which came and went, few European travellers passed through requiring accommodation, and the townspeople ignored Dicky's efforts to promote a New Plymouth hotel. Ngamotu was hardly a place they could drop into, and they now had the Devon Hotel, the Seven Stars Inn and two public houses in the town to serve the slowly growing demand. Anyway, with so few paying jobs, men did not have the wherewithal to splurge on ale and spirits.

To ensure continuing church attendance, Creed had taken over a little, partially built sandstone chapel in the town and now *he* walked the 3 kilometres on Sundays, instead of the worshippers. Barrett was unable to make the same arrangements for his hoped-for hotel customers.

Apart from its taverns, the town was developing in other ways. It now boasted several wooden and 120 cob and raupo houses, four wholesale and retail stores, and a Horticultural Society. The Huatoki and Henui streams were bridged, and a ferry service across the Waiwakaiho River served the outlying farms. But the town was in low spirits, despite this evidence of progress. There were robberies and alcoholism and husband and wife dramas, indicating the difficulties under which people were living.

This was the New Plymouth Barrett left in the first week of February 1843, when he was finally in Wellington to give evidence before Spain. By then, the Land Commissioner's Court had been meandering on for nine months in a decidedly slow but thorough way. Barrett was one of the last witnesses to appear, and now had a chance to meet the man who would adjudicate on the land deals that he and Wakefield had engineered, who was, in George Clarke jnr's words,

> ... of solid intelligence, but with a good deal of legal pedantry ... slow in thinking ... wooden in apprehension ... steady and plodding ... thoroughly honest in intention, and utterly immovable to threats, though he might have been softened by flattery.[14]

Spain had been briefed that the usual rules required in an English court of law may have to be dispensed with or adjusted in New Zealand, and that he would have to look for 'real justice' using his intuition, without regard for the usual legal processes. He was advised simply to examine the best evidence he could procure. Spain had taken an interest in learning Maori laws relating to land, after accepting the brief, but his grasp was superficial. He was not prepared for the intricacies inherent in adjudicating on two different cultures' values; but he did understand how precious the land was, both to the tribes and to the innocent settlers who had purchased in good faith from the New Zealand Company.

On that company's behalf William Wakefield, Dr Evans, Halswell, Hanson, and others cross-examined witnesses, while George Clarke jnr was appearing as advocate for the Maori. Only nineteen at this stage,

Clarke was a tall, gangly, gaunt youth, wearing a frock coat for his appearances. Some people sensed his appointment to this court as nepotism, since his father was the Protector of Aborigines, but it was Hobson, not his father, who had appointed him, on the recommendation of the Chief Justice, the Attorney General and others.

Clarke had been born and bred on a missionary station in the Bay of Islands, spoke the Maori language as well as any European at that time, and was *au fait* with Maori mores and customs. He was perceptive and well balanced for his age, aware of the reasons for the prejudices he encountered, was not overawed and was incisive in some of his court work. However, the more sophisticated members of the Wellington society snubbed and jeered at Clarke's youth, preferring that Halswell or Thompson (the magistrate at Nelson) — both from English universities and members of the English Bar — should be doing his job.

Wakefield and the company's men gave Spain little co-operation. In a concerted campaign against him, they laughed at the Treaty of Waitangi and its concessions to the Maori, and were irked by this examination of their purchases and questioning of their titles. Jerningham Wakefield openly interrupted proceedings with rude and angry outbursts.

Their personalities were familiar to Barrett, who underwent a marathon grilling on Wednesday 8 February 1843. This first session in the witness box lasted from 10 a.m. until 8 p.m., and subsequent sessions continued on and off until 16 February. Edward Meurant, the Wesleyan whom Barrett had last seen buying land for his church's mission in 1840 at Ngamotu, was the court interpreter. Colonel Wakefield appeared briefly. He asked Barrett to recall the circumstances which had led to the buying of Whanganui-a-Tara, then stepped aside and let Dr Evans take over the cross-examination.

Barrett gave a fairly full account of the week's events in 1839, when Whanganui-a-Tara had been bought. He stated that he and the colonel had visited Ngauranga, Kaiwharawhara, Pipitea, Kumutoto, Waipero and Te Aro in one day, and that those people's reaction was 'they could not give their consent to sell the land before they had seen the principal Chiefs in

Port Nicholson, viz., Warepouri [sic], Te Puni and Matangi, that was all.'[15]

This could be an example of Barrett not understanding accurately. Perhaps what those chiefs actually said was that they would not give their consent to sell and that Wakefield had better see the principal chiefs Te Wharepouri, Te Puni and Matangi if he wanted land; because it was the only occasion when Barrett's interpretation of the anti-sellers' reply was not tersely negative. It may have been an encouraging enough loophole to give Wakefield the impression that those chiefs could be manipulated. According to Barrett's later evidence, when he had informed Wakefield that the people of Te Aro and Pipitea were unwilling sellers, the colonel had not seemed unduly put out: '... they would very likely make up their minds before the expiration of the week,' he had replied.[16]

This conviction of Wakefield's may have been strengthened by Barrett's testimony that, in the 'next day's' discussions, the Maori asked Barrett to

> make haste and buy the land so as the white people could ... stop the natives from going to war among themselves — they asked me if there was plenty of ammunition on board ... for ... they expected the Ngatiraukawas [sic] down on them ... I told them that when they had all made their minds up and agreed to sell the land — that they would have the Trade on shore ...[17]

The court interpreter's hand-written records in National Archives show how confusing all the evidence was, and how unskilled in cross-examinations were the people involved. Barrett gave an ambivalent 'No' when Evans asked him, 'Did you then say to him [Wi Tako] that the payment was for the land or the anchorage?', and 'No' again when Evans continued, 'Did you ever say so to Taringakuri or others', which did not answer the questions;[18] but Evans apparently failed to challenge him and Barrett continued to give his honest, sometimes contradictory answers in reply to earnest, occasionally awkward questions.

He had not been called by Wakefield as a witness, when the Court first opened. Clarke had written to his father, five months after the proceedings started, saying that the company was 'afraid to produce him', but that Barrett was 'not such a rascal as to perjure himself . . . and . . . will after a good cross examination expose all their vile proceedings. . . .'[19]

In his questioning Evans returned to Te Aro pa's problems.

Evans: 'Do you remember going to Te Aro Pa on Sunday morning with Colonel Wakefield and Dr Evans?'
Barrett: 'Yes. We went ashore and into the pa where the natives complained to Colonel Wakefield that they had never received payment for their land and had never sold it. We walked about the pa conversing with the natives and then left.'
Evans: 'Was any bargain made or anything said to satisfy them?'
Barrett: 'Nothing. Other than Colonel Wakefield told me to tell them he would send them blankets.'
Evans: 'Was anything said about what they were to do in return for those blankets?'
Barrett: 'Yes. It was said the blankets were for the people who had received no payment for the land.'
Evans: 'Did the natives agree to take those blankets for the land?
Barrett: 'They did not.'[20]

It seems that Evans, unsuspectingly, had elicited some of Clarke's hoped for 'vile proceedings' in open court.

Clarke attempted to squeeze more of these 'vile proceedings' out of Barrett when he took over the questioning. Barrett stated openly that he was a paid servant of the New Zealand Company in these transactions, and that he had known that some chiefs were not willing to sell their land. Clarke handed Barrett the Deed, and asked him to explain it to the court in the Maori language, in the same way he had on board the *Tory* to the assembled chiefs. Again Barrett was put on the spot by the legalese that Jerningham Wakefield had so lovingly inscribed on the parchment. Again, he stumbled through in Maori, which was retranslated into English by Meurant:

> Listen Natives, all the people of Port Nicholson, this is a paper respecting the purchasing of land of yours, this paper has the names of the places of Port Nicholson understand this is a good book. Listen the whole of you Natives — to write your names in this Book — and the names of the places — are Tararua continuing onto the other side of Port Nicholson to the name of Parangarahau; it is a book of the names of the channels and the woods, the whole of them to write in this book people and children the land to Wairaureki, when the people arrive from England they will show you your part, the whole of you.[21]

Barrett's interpretation of the 1,600-word deed was conveyed in a mere 115 words. Then Clarke had some questions:

Clarke: 'Did any of the natives express unwillingness or require an explanation?'

Barrett: 'Some did not like to sign their names. . . . I told them they could not take an article ashore until they had sold their land and signed.'

Clarke: 'Did you tell Te Aro and Pipitea Maori not willing to sell that the Europeans did not like the land and only wanted an anchorage?'

Barrett: 'No. But I will state what I did tell them. Some natives came to me and the Colonel and asked if it was a good place. I said Yes, it was a good harbour. The natives said the land is bad, and I made mention "Never mind the land . . ." as long as there is a good harbour.'

Clarke: 'Repeat the last sentence in Maori.'

Barrett: 'Hawatu nga maunga kai te pai te awa e turanga mo nga kaipuke.'

Court Interpreter:

'What consequence about the mountains the harbour is good.'

Clarke: 'Will you swear that you never told the natives that the reason you wished them to sign the deed was that their names might be seen by the Queen and that the Queen might know they were chiefs?'

Barrett: 'I never said that I wished them to sign their names to the deed that the Queen might send them out presents — I said that when they signed their names the gentlemen in England who had sent out the trade might know who were the chiefs.'[22]

More questioning by Clarke elicited one grey area not covered by Barrett. The chiefs had not specifically indicated that they wanted their pa, cultivations and urupa reserved for them, but neither had Barrett specifically stated that these would be included in any sale to the New Zealand Company.

On and on it went for Barrett during that testing week. Describing Barrett later, Clarke was slightly kinder than Jerningham had been of Clarke: 'a decent fellow enough among men of his class',[23] but very ignorant. He ventured that Barrett had not clearly explained the system of reserves to the Maori, with which Barrett would have agreed, and that he did not even understand the English meaning of the deeds he professed to interpret, with which Barrett would have agreed even more wholeheartedly. Clarke illustrated this by asking Barrett to repeat in Maori, as closely as he could, what he had said to Wi Tako and Wairarapa to make them understand that, by signing the deed on the *Tory*, they were indicating their willingness to sell their land to Wakefield. All along they had denied that Tiki had told them this. Barrett's effort, when translated back into English by the court interpreter, amounted to 'This is the payment for the land in the Vessel — it is written in the paper [the Deed].'[24]

Clarke's description of Barrett's whaler Maori, that it bore 'much the same relation to the real language of the Maoris as the pigeon English of the Chinese does to our mother tongue',[25] is probably not far off the mark; but though Barrett could not speak Maori fluently, he could understand well enough to give an adequate translation when required. Wakefield and Jerningham Wakefield's reports, about the long negotiations in Whanganui-a-Tara and the speeches made by several chiefs, demonstrate that Barrett gave full interpretations to the two men. Either that, or he or the two Wakefield's had inventive imaginations.

Towards the end of Barrett's evidence, Spain had the document produced that Barrett and Reihana Rewiti (Richard Davis, the turncoat mihinare) had composed between them after the disturbance over Revans's house, and which the chiefs at Te Aro had signed. He asked Barrett whether he and Reihana had 'made the natives understand they were signing away their land to Mr Shortland on behalf of the Queen so that the Government might fairly compensate them for the land that had not been fairly purchased.'[26] Barrett said that they had. Spain asked Barrett to tell the court in Maori what the document said. Barrett obliged, then the court interpreter translated his effort:

> The whole of the people of the Pa of Taranaki speaking to Shortland to the Queen Victoria, the land in the Channel of Port Nicholson in New Zealand. The . . . [obscured] of the white people. Mr Shortland says he will judge the talk about the land of the Governor of NZ. if they had not paid for the lands he the Governor will settle the talk.[27]

The next day, Reihana was sworn in and, to Spain's query, said that he recognized the document which Barrett had translated the day before, and that he and Tiki had translated it to the chiefs. Spain then asked Reihana if he and Barrett had 'made the natives understand, etc.' Reihana replied that he did make the chiefs at Te Aro understand; he told them that when the Judge (Spain) came down he would settle it, and that they were satisfied. Later in his evidence, Reihana added,

> Mr Shortland asked if I would ask them if they had been paid for their lands. They said the only things they had received were presents which Tako had brought them. Mr Shortland asked me to ask them what was to be done about their lands. They said the land was to be left for the natives in the hands of the Government to look after it.

Mr Shortland said it would be well to write it down, that they consented for the Government to look after the land.[28]

The result was, of course, the document which Barrett had translated the day before. Spain then asked Reihana whether he and Barrett had made the chiefs understand the contents of the document. Reihana said, 'Yes, it was repeated several times over to them.'[29]

Apart from the legal obfuscation in the evidence, the incident once again brings into question the extent to which Barrett and Reihana had made themselves understood when they spoke to the chiefs on these territorial matters. Reihana testified that he, as well as Tiki Parete, took pains to make the situation clear to Te Aro; but what was emerging was that Reihana's English was as limited as Barrett's Maori. The New Zealand Company had subscribed to trouble by expecting its Principal Agent to arrange land transfers which would satisfy both sides without interpreters of the highest quality. The competence of the court interpreter also comes into question.

Following Barrett's evidence, which had left everyone in no doubt that Wakefield had ridden roughshod over the Te Aro chiefs' definite refusal to sell, Spain had Te Aro, Kumutoto and Tiakiwai chiefs appear before him in the last days of the court's sitting. They agreed at last to accept payment for their land, as partially arranged with Shortland in 1840, as well as retaining possession of their pa, cultivations and urupa.

An icily polite correspondence ensued between the negotiators of the amount to be paid — Clarke for the Maori and Wakefield representing New Zealand Company interests — until the Principal Agent finally achieved supposed legal possession. It came at a further crippling cost for the company and damage to the colonel's reputation. In his report to the British Government, Spain described Wakefield as being undeviatingly unco-operative and obstructive, characteristics which he maintained in later hearings over land at Wanganui and Manawatu.

Before he left town, Barrett put an advertisement in the *Gazette* for extra hands for the coming whaling season, to replace those who had drifted away from his disconsolate establishment. Then he joined up with Cooke (who had so firmly 'settled' the Waitara dispute for Wicksteed), and started out on 20 February 1843, to help Cooke and two or three stockmen shepherd 'seventy head of cattle and a large flock of sheep' back to New Plymouth. Charles and John Hursthouse, two settlers heading for that town, joined them just out of Waikanae, on a gruelling trip that eventually took well over a month. They travelled on foot with that considerable mob through bush and scrub, over sandhills and along difficult, rocky shoreline, and in inclement weather; but Maori on the route, many of whom had never seen a 'kaukau' before, helpfully lent canoes for the men's river crossings, while the animals swam.

They reached Wanganui after eight days and stopped for two nights' rest. Barrett introduced his companions to Bell, an old Scotsman there, 'celebrated for hard work talking drinking and swearing . . .' Bell gave them 'two meals in 3 hours, talked incessantly all the while', plied them liberally with whisky, and 'in enforcing his arguments knocked Cook (*sic*) and Barrett his supporters on either side black & blue!'[30]

Charles Nairn joined them at this time. He had just completed supervising the cutting of the road from Taranaki to Wanganui, but, even with his help, moving all the livestock across the river took a good twelve hours. Five days later, on 8 March, when the weather broke, Barrett had to remain with the muster while Cooke went ahead looking for shelter. The heavy rain and broken country took its toll and the Hursthouses turned back to Wellington soon after. However, the drive was time well spent for Barrett, who had seen little milk, butter and cheese for a decade. Now there would be cows and sheep available for his farm, and fresh beef and mutton would be welcome additions to the salted meats, fish, rats and dog, which he and his dependents had become used to.

The whalers who had answered Barrett's advertisement, and the whaling gear, financed by Rundle, which Barrett had ordered in Wellington, arrived at Ngamotu from Port Nicholson at the beginning of May, ready for the first influx of the mammals; but although plenty of whales swam into view at the start of the season, none were caught. Even Barrett's new men had quickly become corrupted by the old hands and turned 'mutinous and incapable'. They were part of a community in depression from hunger and isolation, moving towards a period of deeper depression. Even on the farming front, caterpillars were attacking the barley, and smut was reducing wheat yields because of the wet summer.

There was a brief period of mutual co-operation between Maori and Pakeha in the community, after an untrue rumour that the common enemy, Waikato, was threatening belligerence again. But mutual distrust returned to a lamentable degree when news came that twenty-two Europeans and four Maori had been killed on 17 June in a stand-off with Te Rauparaha over land at Wairau.[31] The confrontation had been exacerbated by government officials at Wairau adopting the armed approach, whereby Wicksteed claimed to have settled the Waitara dispute effectively several months before.

Also in June, the funds of the Plymouth Company were exhausted; and, under orders from the New Zealand Company directors, Wicksteed began cutting back savagely. He reduced the labourers' wages by more than half, issued no more stores or rations, dismissed Cutfield, who had taken on the position of Company storekeeper — and took over that job himself. He informed Carrington, who had with reason been feeling increasingly insecure about the merger of the Plymouth and New Zealand companies, that his services as Chief Surveyor of the Company would cease on 31 March 1844. Subsequently, within a short time, Carrington's work team had reduced in number from eighty to two or three men, his requests for very necessary equipment went unanswered, his salary ceased and, eventually, he was told to stop surveying altogether. Since his remuneration was gauged on the number of sections he produced, he was going to be greatly out of pocket.

Carrington did not wait until March. He began preparing to leave as soon as possible. He saw himself and Barrett as joint sufferers, as a result of their connection with the company and, over the next three months, they grew closer. Carrington's journal was sprinkled with entries like, 'I go to Barretts with Willie [his brother Wellington], Ockey [Octavius]

& Halse . . .'; 'Mr. Barrett at Ockeys in evening talks about Waikatos . . .'; 'Barrett called . . .'; and Barrett gave Carrington a farewell gift of a toki.

They began to talk confidentially in some depth. Barrett was having second thoughts about his behaviour over Toponunga and the Waitara land.[32] In a day of catharsis, he confessed to Carrington, in front of other witnesses, that Toponunga 'never received any payment for his land; that he never was a slave; that most of his near relations never sold their land, never put their name to paper, and never received any payment from the Company', confirming completely what that chief and others had told Carrington.[33]

And the two men swapped documents, in case they needed some redress should the company 'sell them short'. Carrington gave Barrett a certificate stating that he believed the piece of level land at 'Whaling Moutaroa' belonged to Barrett. In return, Barrett gave the surveyor Wakefield's letter, written in November 1839, which declared that Wakefield had bought and paid for land on the west coast of Te Ika a Maui, including Taranaki. It was Carrington's protection against 'any claim of wrongdoing and to force compensation from the company for his dismissal', since he reckoned that Wakefield was prevaricating — he had not properly 'purchased' north Taranaki until Barrett had done so on the New Zealand Company's behalf, between December 1839 and February 1840.

Carrington and his wife and their five children — they now had two more New Zealand daughters — sailed on 27 August 1843, and Barrett fired a poignant three-gun salute of farewell as their ship moved off. Barrett could not sail away. When the New Zealand Company settlers, who represented a high proportion of the tax-payers in New Zealand, were unable to scrape together their tax payments, the New Zealand government became indebted to the Home government for its own dues. It could not even afford the salaries of its own officials. Those uneasy bedfellows, the New Zealand Company and the government, were now tucked up and declining together, taking the community at New Plymouth with them, and Barrett was part of the downward slide.

Chapter 14

SEPTEMBER 1843–23 FEBRUARY 1847
ON THE STAND AND LAST DAYS

In the sombre environment of community and New Zealand Company stringencies, Barrett struggled on. Contrary to what is written in the Barrett Journal and perpetuated in his legend, Dicky Barrett did not go on and prosper. The Barretts never built a substantial home at Ngamotu, like their one at Te Awaiti. They continued to live in a simple whare-type dwelling, surrounded by 14 acres of land of questionable ownership, and farming more land nearby. By 1843, Caroline was about fourteen, the age at which girls ceased any formal education. She would have been helping her mother and father with household and farming chores; Sarah, aged eight, would still have been at the school next door, run by Mrs Creed for the children of the whalers. Mrs Creed had no living children of her own, despite being continually pregnant.[1] Barrett's New Plymouth hotel remained very basic and was not included in the town's records of hostelries and Dicky's ambitious plans went no further.

The Wairau affair hung like a constant cloud over everyone. Shortland, who had taken over as 'Officer administering the Government' in Auckland when Hobson died, had further antagonized the settlers by describing the actions of the European officials at Wairau as typical of the New Zealand Company's habit of acting 'in open defiance of the Government'. When he uncompromisingly declared the whites' actions there illegal, Te Ati Awa living around Nelson melted away, rather than wait for te riri Pakeha (the white man's anger), which they had learned to expect if the Pakeha had a decision given against them. But those who returned to New Plymouth, although finding some amiable interaction on a personal level, encountered general te riri Pakeha anyway, inflamed by Shortland. After Wairau, he had forbidden settlers

*New Plymouth in 1844, painted by Edwin Harris, from the junction of Queen and St Aubyn Streets. Pukeariki (Mount Eliot), later completely levelled, is in the left foreground. Pukaka (Marsland Hill), levelled on the top in the 1850-60s for a military fort during the Taranaki Wars, is in the left middleground.* TARANAKI MUSEUM

anywhere to move onto land before their titles had been decided by Spain, so upon hearing this Te Ati Awa clung to land which the Europeans saw as belonging to them as part of Barrett's 'purchase'. Insecure on their scattered sections anyway, the settlers around New Plymouth wished more than ever that they had never left their home country. Te Ati Awa may well have said Amen to that.

In July, there had been rumours again of impending invasion from Waikato, despite Hobson's payment to Te Wherowhero. No taua showed, but Wicksteed swore in a large number of Europeans as special constables, as a precaution. Puketapu, Ngati Te Whiti, Ngati Tawhirikura and other Ati Awa hapu saw these constables as disguised blackmailers, as the settlers moved purposefully onto some, then more Waitara land, defying Shortland's decree. United by this infiltration of the prized fertile soil, whanau combined to deny any Pakeha rights of occupation, whether it was their hapu home ground or not. They implemented a campaign of non-violent protest, dismantling new fences the settlers built and preventing the movement of stock; and when the double-barrelled guns Tiki had promised them finally turned up for distribution, they refused to take delivery.[2] The previous gifts they had received from his hands had led to the Europeans seeming to expect rights to any Ati Awa land.

In the largest demonstration, nearly 200 Ati Awa from kainga near the Barretts through to Waitara and beyond, sat unarmed on the road to Waitara, to impede the cutting of survey lines. '. . . you are all Wicksteed's slaves, and we will not listen to you,' they jeered (mimicking what had been said to them) when Ockey Carrington attempted to reason with them.[3] Wicksteed, who was finding himself beset by problems and unpopularity on all sides, ordered the surveyors to stop work, horrified at the thought of another Wairau. By September, a group of irate townspeople, quite ignoring their own violation of Shortland's directive, drew up a document to send to him about the Maori resistance to the 'quiet occupation' of the land by the whites. They did not appreciate that the Maori, by their passive resistance, were quietly showing both their grievance and their desire for peace.

Well away from this daily challenge to the land Barrett had acquired, his 1843 whale record was as dismal as the year before, despite the appearance of a record number of whales in the roadstead. Barrett's whalers and their families still lived cooped up in the barracks-like, large whare, which had greeted the immigrants two years previously; but, despite their lack of success and discipline, Barrett was too soft-hearted to put his men off; they would get no work with Wicksteed — he had too many labourers on company pay already. So the whalers drank away their time and prospective meagre lays and three of them died that year as a result of their intemperate behaviour. These stresses and the constant food shortages inevitably affected Barrett, and may well have slimmed his formerly rotund girth down to a finer silhouette, as depicted in the only known portrait alleged to be that of Barrett.[4]

In May 1844, Spain finally arrived in New Plymouth, to adjudicate on the ownership of the land Barrett had bargained for. The Land Commissioner travelled overland to New Plymouth with, among others, George Clarke jnr and William Wakefield. In Wellington, the latter had impeded and delayed Spain's court sittings with arrogant and wilful absences, reputedly to give New Zealand Company directors in England the opportunity to stop the inquiries. However, on this trip, Wakefield had apparently chosen to use his proverbial charm on the commissioner.

As a result, Spain opened his New Plymouth court with a bombshell, dashing Te Ati Awa's hopes with his opening pronouncements. He said that both he and Wakefield agreed that the former owners of the land had lost all rights to it when they became slaves of Waikato; and that, further, in his view, the Governor had bought land rights to Taranaki from Te Wherowhero and Tainui, who had won them by right of conquest. He went on to impose pre-emptive conditions under which he would hold his investigation: only present residents would have their land rights at Taranaki even considered and, even more presumptive and in denial of basic Maori law, chiefs would not be able to claim rights to land in two places at once. Would he have therefore disinherited the settlers from their property in England? Wakefield followed this up by declaring he would forego any claim on the land south of the Sugar Loaves, which Barrett had 'bought' from Taranaki iwi.[5] This was possibly a concession given in return for Spain's opening statements. Barrett's Ngamotu deed therefore became Spain's focus.

With these restrictive terms in force, Barrett took the stand on 31 May, in front of about a hundred settlers and an unknown number of Te Ati Awa, many of whom had been trickling north for this case involving their land. Barrett was the first and major European witness in New Plymouth, instead of one of the last, as in Wellington; and it appears that any leanings he may have had towards acting in his previous role as Agent for the Natives, had vanished. He was acting as agent for himself, and, perhaps, slightly for Wakefield. Despite his earlier private admissions to Carrington and Toponunga, he stuck to the New Zealand Company line in his evidence — that he had truly bought all the land described in the deed.

Initially, Barrett gave a factual description of the 'buying process' as he saw it, including his admission that he could not explain the system of reserves to the chiefs, although one had understood him. Dorset had recounted in Spain's Wellington court how one man said that, 'now he had a reserve of Land he would be able to get a white rangatira husband' for his daughter!'[6] It emerged for the first time that the final bargaining negotiations took place on the *Guide*, with only a few Ati Awa present.

Then, in contrast to his frankness in the Wellington court the year before, Barrett began to alternate between candour and defensive prevarication. At this stage he was able to avoid the Toponunga issue with his questioners, Spain and Clarke:

Q 'When you agreed with the natives to make this purchase, were any natives residing at Waitara?'
A 'No, there were not. I do not think there were any residing at Waitara at that time.'
Q 'Do you know where the natives now inhabiting at Waitara were living at that time?'
A 'Some were at Waikato, some at Waikanae, and some at Queen Charlotte's Sound.'
Q 'Was not this district nearly deserted at the time of the arrival of the *Tory*, in consequence of the threats of the Waikato people?'
A 'Yes, it was.'

A double question then enabled Barrett to avoid giving a straight answer:

[Q] 'Were they [the tribespeople at Ngamotu with whom Barrett had negotiated] the real proprietors of the land? Or part of the tribes which formerly occupied it?'
[A] 'I believe they were, partly, by relationship.'

Then the prevaricating started:

[Q] 'Do you think they [the New Zealand Company] bought the land with the consent of the majority of the natives residing on the land sold?'
[A] 'Yes.'
[Q] 'Do you believe that they were the proper persons who had the right to sell the land?'
[A] 'I believe they were. . . .'

[Q] 'Do you know whether, by the native custom, those natives who formerly resided here, but who had left the place, had forfeited their right to the land?'
[A] 'According to my opinion, when they were driven away by the Waikato, being absent eight years, they had forfeited their right to the land.'[7]

Barrett made no effort to right his own wrong. Instead, he pleased the whites and his connections among the 'remnants', who had kept the tribe's land ahi ka; but he infuriated those who had left Ngamotu with him on the heke, or who had fought in Te Rauparaha's campaigns, or who had left temporarily at other times. He also, apparently unthinkingly, put his and Rawinia's land rights in jeopardy, since she had been away from Ngamotu for eight years.

Three moderately minor details about the Ngamotu deed emerged while Barrett was being cross-examined. Firstly, Barrett did not mention the fracas between Taranaki and Te Ati Awa in his evidence. Secondly, it became apparent that Waiaweka[8] had illegally signed both deeds; and thirdly, Ati Awa had sent about £10 or £12's worth of the payment on the *Guide* to relatives at Whanganui-a-Tara. This gave some rebuttal to those Ati Awa who had returned during the past few years and claimed that the remnants had cheated on them by keeping all the payment for themselves, because Tuarau was supposed to have delivered a portion of the purchase goods to them at Port Nicholson.

At Clarke's request, Barrett went through the more than forty signatures on the deed and identified them, one by one, as either from Ngamotu or Puketapu hapu of Te Ati Awa; however, even the Puketapu signatories were flying in the face of their hapu's wishes. The only excuse Barrett may have had for thinking they could sell the land right up past Parininihi is that, during the fifteen years he had been in New Zealand, so many of them may have mentioned connections with hapu living within those boundaries.

Barrett swore his evidence and, after the weekend break, Spain had it translated into Maori and read out by Forsaith, the court interpreter[9], to the assembled Maori. Spain offered them sincere help, after asking them to respond to it, which message he then spoilt by launching into a condescending homily, implying that jealousy and conflicting motives might influence them. He finished with, 'A lie will not only subject you to the scorn and contempt of your fellow men, but is wicked and offensive in the sight of Almighty God.'[10] There was a significant pause, before one chief said, '. . . a great deal of what Barrett said is right; some is not.'

With Barrett, their main link to the European court, not coming out with evidence to support them, and the Commissioner having pre-empted their cause, Te Ati Awa in the courtroom chose taciturnity and evasion. No one spoke representing Waitara hapu or the other areas that were starting to become subjects of hot dispute.

Clarke, appearing on behalf of Te Ati Awa, had two aims. He wanted to elicit replies from witnesses which would, firstly, clarify for Spain the rules of Maori land ownership and, secondly, show that Spain's initial pronouncement, as he opened his court, were wrong in law: that Te Ati Awa, who had moved away from Taranaki, had in no way lost their rights to their tribal land. To achieve this, Clarke asked the few Maori who came forward the same questions, but he had little success. They maintained their reserve or obfuscated; but Spain did not see the significance of their silences as showing neither agreement nor acceptance. To one query, however, they all opened up — Te Ati Awa had moved away from their tribal lands of their own free will, not because they had been defeated, and Clarke must have thought he had partially achieved his purpose. Following that, most were vague about whether Tiki had read out or explained the deed, and one or two suggested he had urged them to sign it by intimating that if they did, 'the gentlemen' in England would know their names. According to Barrett's evidence, he had told them '. . . let the gentlemen in England see they had sold their Land.'[11]

Only two gave direct information. John Ngamotu detailed the areas sold erroneously: Waitara, Pukerangiora, Wakahingo, Ngati Maru, Parininihi, and many others belonging to other hapu or iwi apart from Ngamotu but were not being lived on at the time. He stressed that Tiki

Parete had not dealt with influential chiefs, many of whom were at Kapiti, Waikanae or the Chathams; that the chiefs who had talked with Tiki on board the *Guide* were not the only owners of the land, were minor chiefs, and did not know the boundaries. He added that Waikato had tried to clear land and plant potatoes, but Te Ati Awa 'remnants' had stopped them.

Following him, Te Mana (also known as Piri Hapimana) corroborated some of the above, said that the land sold was only at Ngamotu, Huatoki, Te Henui and Waiongana, yet supported other Barrett evidence. Then he let the cat out of the bag, saying that the Maori had decided to gang up against Barrett, about hearing the deed translated, because the payment was small and they had not received double-barrelled guns.

Next, Clarke called and questioned his own witness — Jacob Wahao, a baptized Christian from Waitara:

Q 'Have you claims on this district?'
A 'Yes.'
Q 'Do you own it? Or the tribe?'
A 'All of us, as men of Ngatiawa.'
Q 'Describe your claim and how it was derived.'
A 'By our ancestor Te Moewakatara, long time ago.'
Q 'Do all Ngatiawa's claims to Waitara come through him?'
A 'Yes.
Q 'Do not each family and individual have boundaried land?'
A 'Yes, the elder brothers, then the younger's.'
Q 'When were these boundaries laid down?'
A 'A long time ago, and as the claimants were successively born, their possessions were marked ... with marks and stones.'
Q 'When did you first see these stones and marks?'
A 'As soon as I could distinguish things.'
Q 'When did Ngatiawa leave this district. What was the cause?'
A 'Twenty years ago. They went of their own accord.'
Q 'Did they go before or after Waikato took Pukerangiora?'
A 'Some time before.'[12]

Under Clarke's persistent questioning, Wahao's replies confirmed what all the other Ati Awa had said, and what Barrett and Love had experienced when trying to establish the trading post at Ngamotu — Te Ati Awa were a migratory people who never lost their rights to their homeland while their tribe inhabited it.

Wahao was uncomfortable with the next line of questioning, which elicited the, for him, disgraceful fact he had been a slave; yet, through this passage, Clarke was able to demonstrate that in Maori custom slavery did not mean forfeiture of tribal-land possession and, after Wahao and about twenty others had been released from slavery by Waikato, he had returned to Waitara and begun cultivating again as of right. He had not received any of the goods given by Barrett for the land. Europeans, moreover, had barged in and sowed wheat on his land. To Clarke's final questions, he said very definitely that he had never heard that 'Te Weoro Weoro [Whero-whero] had sold 'this very land' to the Governor, the late Captain Hobson. Clarke rested his case there.

On the last day, Colonel Wakefield recalled Barrett for further questioning:

Q 'Do you know the native, named Mana, who was called yesterday?'
A 'Yes.'
Q 'What character does he bear among the natives of this place?'
A 'A very good one....'
Q 'State to the Court what he said yesterday to the natives, before he was examined.'
A 'He said he would tell the Commissioner one word which the others wished to keep back. He said to me that the natives would not say that they had heard the Deed read over before they signed their names; and he would speak the truth.'
Q '...You have heard several native witnesses deny that they heard the Deed read over and explained by you, before it was executed. Do you still adhere to your former evidence on this point, and affirm that you did read over and explain it to them, before it was signed?'

A 'Yes, I did, as well as I was able.'
Q 'And do you believe that they fully understood at the time that they were disposing of the land, between the Sugar Loaves and the Taniwa? . . .'
A 'I was living among them between 2 and three months before the sale, talking to them constantly about the land; and I believe they knew very well what they were about.'
Q 'Do you think they fully understood what payment they were to receive?'
A 'Yes, they saw it all laid out on the deck of the *Guide*; and they desired me to send it ashore.' [Only a proportion of Ati Awa were on the *Guide*.]
Q 'And that they were satisfied with the payment they received?'
A 'Yes. There is one thing I would mention again. I promised them, when they asked me that they should have some double-barrelled guns, I promised them at some time or other they should have a case of these guns.'
Q 'Did the natives complain to you of the sale within a short time afterwards?'
A 'I sailed for Port Nicholson the day after the Deed was executed.'[13]

Spain, honest, indefatigable lawyer though he was, was running out of patience, after thirteen months of this exotic law-court work. He made no effort to seek more witnesses or check the evidence and closed the court after just one week's sitting.

Two days later, he presented his findings. He declared that Te Ati Awa had rights to their land at Port Nicholson, because they had cultivated the land there for ten or eleven years, but — with blatant disregard for Maori evidence given in court, and tribal land law — no rights now in Taranaki, because they had not come back in time to establish their case. With that, he awarded the New Zealand Company about 60,000 acres (24,000 hectares) of Ati Awa land, from just north of the Sugar Loaves to the river at Taniwha. He also awarded the Wesleyan Mission 100 acres (40 hectares), and Barrett 180 acres (73 hectares) at Ngamotu, 'through the goodwill and given word of Wakefield', consisting of two lots: 80 acres (32 hectares) in section 23, and 100 acres (20 hectares) in section 37 (see pages 248–49); but he said that the Barretts were not to receive title outright. The land was to be part of the tribal grant, and would be held in trust by the trustees of the native reserves. In his concluding comments, Spain said:

> . . . It affords me great satisfaction to be able to say that Barrett's testimony in this and other cases where I have had occasion to examine him, corroborated as it has substantially been by Native testimony, has led me to the conclusion that he has told the plain honest truth as to what took place in all the transactions . . . and he has not hesitated to do this when his testimony has gone in other cases as much against his Employers, as it has done in the present instance, in their favour.[14]

He added that Barrett's 'purchase' had been free from the 'carelessness' associated with the company's other negotiations.

Te Ati Awa in New Plymouth were left with 6,000 acres (2430 hectares) of reserves, being one tenth of the company's total acreage in New Plymouth — in line with the New Zealand Company's now established practice — as well as their pa, cultivations and burial grounds. Spain also promised them reimbursement of £200 in other ways, instead of the double-barrelled guns, which had been lost in the *Jewess*. If they were now given guns, Spain professed to have been advised, Waikato may invade to claim them; and if he gave their value in money, it would be difficult to divide up among the very many claimants. He decreed, therefore, that the New Zealand Company should instead build men's and women's hospital wards, to the value of the £200, for the good of Te Ati Awa, to be sited upon the native reserve in the town. He went on to tell the Maori how much they were benefiting from having the Europeans: there were no more wars, they had civilized clothes, money, and English labour opportunities, farming methods and general industry, and leisure to be taught the way to worship Almighty God, all guaranteed to lead to '. . . a life here to ensure your happiness hereafter'. The minutes detailing his judgement end with the sentence, 'The Court then broke up.'[15]

Indeed! There was an immediate uproar among the Puketapu hapu of Te Ati Awa and Te Ati Awa in general, excepting Ngamotu hapu, and any euphoria Barrett may have felt at Spain's grant and commendations would soon have been lost. The affected chiefs drew together to write strong protestations to the new Governor.[16] Wiremu Kingi Te Rangitake, prominent Ati Awa chief who had travelled to New Plymouth from Waikanae (where he had lived since emigrating in the 1820s) for the hearing, wrote on behalf of these irate chiefs, with Clarke's backing. He claimed strongly they had never left their land, nor had they alienated it in any way. They did not want the settlers to leave; they just wanted them not to settle in certain of their favourite localities — to whit, Waitara. He challenged FitzRoy — did he not love England, 'the land of your fathers, as we also love our land at Waitara?' 'Waitara shall not be given up' became his catch cry.[17] He suggested that there were vast forests, in the area outlined in Barrett's deed, which the settlers were very welcome to fell and clear and live on. Clarke sent a letter to the Governor supporting this missive and warning '. . . If the Govt are determined to put the settlers in possession of their lands . . . they must do it at the point of a bayonet . . .'[18] Te Ati Awa did not wait for Clarke's or their letter to take effect. They occupied their favourite places and began to remove settlers' belongings.

Not only Te Ati Awa were outraged. Elsewhere, people criticized Spain for accepting the Ngamotu deed as valid. Barrett had had it signed on 15 February 1840, but the New Zealand Company and the British Government had agreed that the company would have rights only to land for which the company had deeds signed before Captain Hobson's arrival as Lieutenant Governor, on 31 January 1840. In defence of it, Wakefield had landed Barrett on 28 November 1839, to buy Taranaki land, expecting to be back in one month; and it could be said that everything had been arranged before Hobson reached New Zealand, and that the deed should have been signed, except for those acts of God — the shipwreck of the *Tory* and the storm that had forced the *Guide* to seek shelter on 2 January 1840. Wakefield himself had indicated that he considered the Ngamotu deed was simply a confirmation of the sale of land, which he had had signed in East Bay; and that he had reconfirmed this only because his instructions were to give the utmost publicity and satisfaction to the natives.[19]

Registering the furore his decision had evoked, Spain stuck figuratively and literally to his guns. He wrote to FitzRoy on 2 July advocating the imposition of a military force and, possibly, a man-of-war, to persuade the Maori that everything was in their best interests; or, as he put it more bluntly, to demonstrate 'our power to enforce obedience to the laws, and of the utter hopelessness of any attempt on their part at resistance . . .'. As far as Spain was concerned, New Zealand had been colonized for philanthropic reasons, 'to benefit the Natives by teaching them the usefulness of habits of industry, and the advantages attendant upon civilization'. The said natives, finding distinct disadvantages arising from their contacts with civilisation, continued to fume, agitate and threaten.

Out of the limelight again, Barrett returned to his farming and prepared for the opening of the whaling season, again backed by Rundle. He must have felt vindicated by the lack of serious trouble in the New Plymouth roadstead. Out of 259 ships which had visited, only one had been wrecked and fourteen anchors been lost. By comparison, Port Nicholson had had far more casualties.[20]

In July, his whalers had the try-pot fires glowing and a satisfactory stench filling the atmosphere, after killing three whales that month, with the added bonus of a large 75-kilogram shark they harpooned, which had cruised in to feed on one of the whale carcases. The resultant shark meat was satisfyingly tasty. But they caught only three more whales in the rest of the season, and when Barrett sent his 20 tuns of oil and 1 ton of whalebone to a Sydney market, the 1844 price for whale oil had fallen drastically.

Meanwhile FitzRoy, although inundated with other unremitting problems elsewhere, pressed Europeans more experienced in New Zealand affairs than himself to join him in New Plymouth to respond to Wiremu Kingi's letter. Bishop Selwyn, the Rev. Mr Whiteley and young Donald McLean, recently appointed sub-protector of Aborigines in Taranaki, accepted and met him there. On 3 August 1844, they conferred with a large number of Maori, Wicksteed, and assorted prominent settlers, in

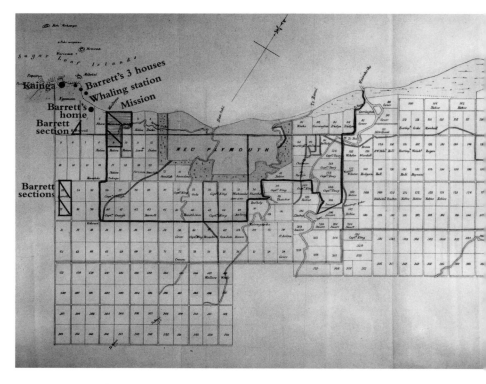

*Barrett's whaling station, his separated sections and the Wesleyan Mission are identified on the top left-hand side of Octavius Carrington's map of New Plymouth. The different awards of land, by Spain, FitzRoy and Grey, are outlined.* TARANAKI MUSEUM

one of Te Ati Awa's important marae, high on Pukeariki.

Like Spain, FitzRoy announced his decision without delay. He would not be confirming Spain's judgement since he could and would not agree that Te Ati Awa had forfeited their rights. He followed this up by offering free passage to the north for any people upset by his pronouncement, and appointed McLean as an independent adjudicator, to avoid any suggestions of bias. McLean was to try to ascertain basically to whom the land belonged, and to investigate the claims of the absentee tribal owners, who had not been in Taranaki for the original bargaining with Barrett. In a whirl, FitzRoy departed after only two days in New Plymouth, to deal with the anti-government defiance of Nga Puhi leader Hone Heke at Kororareka, as well as with news that the New Zealand Company had folded financially. Twenty-four year-old McLean remained. He talked with chiefs all over the district, realized the splinters

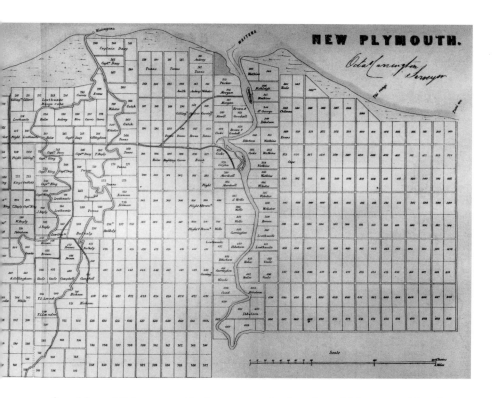

of problems Waitara would always contain, but found Ngamotu Maori still willing to sell 3,500 acres, as long as their cultivations would be reserved for them. Of the now nearly 100 adults living around Paritutu, eighty-four signed this agreement with him.

He took three months to make his report to the Governor, who returned to New Plymouth to announce, on 8 November 1844, McLean's recommendation that nearly the whole settlement must go back to Maori ownership. FitzRoy knew the suffering this would involve, so he stipulated that if the company paid £350 extra to Ati Awa, he would issue a Crown Grant for the area where most of the settlers were living. His ordinance meant that only 3,500 acres of Barrett's original 'purchase', right up to the White Cliffs and down into the Taranaki tribe's country, remained in the New Zealand Company control. He allowed the Barretts to retain two sections, but with an appreciable gap between them, while other farmers, in outlying districts, were offered Carrington's conscientiously and artistically designed green belt of parks, cemeteries and native reserves.

FitzRoy again denied emphatically that Te Ati Awa absent through slavery or evacuation had lost their land rights. Imagine the rumpus, he suggested, which would have erupted had British prisoners-of-war returned to England to find that their estates were forfeited. He blamed Barrett, an 'interpreter incapable of translation', who had dealt with only the forty or so people at Ngamotu at the time, who owned about one-thirtieth of the land. FitzRoy reiterated Clarke's warning, saying that it 'was plain that Spain's award could not be implemented without bloodshed, and the probable ruin of the settlement'. He again refused to sanction it.

Somehow or other, the New Zealand Company men in New Zealand obediently scraped together the collateral, and the Maori were given their payment: blankets, shirts, tobacco, etc., and £50 in gold to make up the value of £350. Then, having already laid off most of their staff in New Plymouth, the directors plaintively asked Her Majesty's Government to rescue them. Some Te Ati Awa were satisfied with their windfall. They did not want to be rid of the Europeans completely, and this solution gave them mana and hope, their land back, and funds which they could invest in stock. Others were disenchanted. They still found Europeans making themselves at home, uninvited and barefaced, on 'waste' land, which was fresh tribal land that Ati Awa intended to cut, dry, burn, dig and sow.

FitzRoy's decision not to honour Spain's rulings meant his high standing with the settlers suffered an astonishingly rapid reversal. As the farmers painfully re-established themselves on the small area of land around the town, there were vituperative requests to Her Majesty's government for the recall of their Governor of 'unsound mind'. Barrett's prestige, too, was declining as steadily as his whaling receipts, his reputation further damaged by FitzRoy's denigratory comments about him. From being congratulated on his land acquisition, Barrett became widely used as the scapegoat, as Europeans absorbed the significance of FitzRoy's decree and met with Ati Awa's staunch resolve not to move from Waitara.

By 1845, Barrett had lost most of his mana. Wakefield was referring to him as 'Dirty Dicky' and the people of New Plymouth ignored him. Apart from blaming him for their difficulties, they no longer needed his knowledge of the waters off the coast, and now there was little need for

*Poharama-Te-Whiti, 1800-1875/76, was described as a staunch friend by most colonists and supported them steadfastly against unfriendly tribes.* TARANAKI MUSEUM

his interpreting help — which had been exposed as unreliable anyway. From being someone whose name appeared not infrequently in immigrants' letters, Barrett almost ceased to be mentioned. His Ngamotu whaling team had its best result ever, but that consisted of a paltry six unprofitable victims. Brown fared little better.

The antipathy against Barrett came not only from the settlers. Some Ngamotu whanau recognized Tiki's contribution to their problems. Wiremu Kingi and Poharama voiced various complaints against 'Barrett's natives', and claimed that Tiki had not paid them for some timber. As their remonstrance, Kingi and Poharama fenced off a village road between his two sections, adding nearly 2 kilometres to the distance Barrett had to transport his produce and drive cattle. McLean stepped into the dispute, but could not persuade the chiefs to back down. Tiki had not paid them, they maintained, and the fence would stay, despite Tiki and Poharama having fought together at Otaka. Perhaps this was a protest against Carrington's subdivision of tribal land; perhaps Poharama did not acknowledge the sections Spain had given Tiki, but Poharama did not interfere with Spain's award to the Wesleyans. Perhaps it was a token protest for the trouble Tiki Parete had brought to the region.

New Zealand Company directors and settlers may have been very short of money, but they were extravagant with their use of paper for wrathful letters to Her Majesty's Government and by October 1845 Governor FitzRoy had received his recall papers. He was replaced by George Grey; but this did not lessen the now endemic tensions in New Plymouth. Maori had returned to live all around the town, and were preventing the passage of Europeans across any of their land, which the settlers had previously used so freely. In the settlers' reaction, muskets and bayonets poured into the town, as they sought to protect themselves against the threats they saw from these increasingly inimical neighbours.

Affairs in Wellington were even more chaotic. By the beginning of 1846, as the tribes in Whanganui-a-Tara reached the end of their peaceful tether with the Pakeha, the Pakeha strengthened their resolve. When Te Rauparaha's nephew, Te Rangihaeata, resisted obdurately a road being built to Porirua, barracks were built in Courteney Street. When Maori reaction against illegal settler occupation in the Hutt Valley culminated in a death at Boulcott's farm in early February, troops were sent to Wellington in response. Wellington sat like a bonfire waiting to be lit.

These events were inevitably bruited round the country and it seems that, like Te Wharepouri before him, as Barrett's efforts produced such negative effects, his health began to deteriorate. Rawinia was apparently also ailing. Wicksteed reported in May 1846 that Barrett had been ill, but was 'quite recovered and making preparations for commencement' of the whaling season; yet by June Wicksteed was writing that 'in consequence of Richard Barrett's severe illness, his party is in a great measure disabled.'21 Barrett's team produced only 5 tuns of oil that season (compared even with Brown's meagre 37 tuns, and, at its end, he had only one boat and seven whalers remaining with him; the rest had drifted away.

As a result, although his farm was 'the only one of any consequence' at Moturoa, according to a census report to McLean in December 1846, Barrett was in even more financial difficulties. He was in debt to Wiremu

Kingi and Poharama, to the New Zealand Company for two horses they had passed over to him, and an unknown one to T.B. Hine, son of one of the New Zealand Company directors, who himself had many liabilities. Whether he still owed money for the hotel lease is unknown.

The settlers cannot have been sympathetic, since they held him responsible for their being marooned with no legal title deeds for their land in New Plymouth. Yet, although Barrett was now *persona non grata*, the essence of his personality was sown in their minds, leading to stories about him in the future that were reminiscent of Barrett's own tall tales. Typical of these was an article about Barrett in the *Taranaki Herald* nearly 100 years later:

> Fight With a Whale Ended In Severe Injuries
>
> '...There came a day when Barrett was killing a whale that had run unusually close inshore. Three boats from Richard Brown's station, which had joined in the chase, hung about within a short distance away. Barrett had four boats, and he captained the seven-oared himself. As a rule these sights were watched from the high ground where the flagstaff stood (Mt. Eliot), but on this occasion, the tide being out and the whale so close in, a group of excited merchants and idlers stood close to the sea watching through one or two telescopes. The men in the boats and all their actions were visible. The whale was spouting blood in a vortex of spume when Dicky Barrett ran in to put the finishing lance into its 'life'.
>
> ' "There She's Fluked!"
>
> 'Suddenly the cry went up: "There she's fluked — by jove, she's done for Barrett." The great tail appeared to fall on the boat, which for the moment was lost to view in the churning cauldron worked up by the dying whale.
>
> 'Soon, however, Dicky Barrett was lifted from her bottom insensible and carried ashore. He lay on the warm sand for some time above high-water mark. Then he was half-carried and half-walked between two strong supporters to Mr. Richard Brown's, which was the nearest house. They were rivals in the whaling

business, but all knew that Mr. Brown would do everything in his power for his friend.

'The men in the boat told the story of the mishap. They said that the whale had not touched the boat, and all the mischief had been effected by the concussion of the air as that terrible tail fell. It seems possible, however, that the tail struck the oars and wrecked the gunwale and the planking. A man was baling hard all the way in to keep the boat afloat.'

*Tarankai Herald*, 29 March 1941[22]

Interestingly enough, there are no contemporary records about this colourful incident. Neither Wicksteed nor Flight nor Newland, whose recordings of day-by-day affairs in New Plymouth cover most items of note at the time, mention the accident. Various items in their pages simply attest to the fact that Barrett was ill and failing for several weeks or months, and was nursed by Rawinia.

Dicky Barrett died on 23 February 1847. One diarist stated that he died of apoplexy; but, given his being overweight for so long, the period of ill-health mentioned the previous year, and the constant stress under which he had lived since 1839, it was probable he suffered a sudden heart attack. His death warranted passing mention in one or two other New Plymouth diaries and in the *Gazette* in Wellington, and Wicksteed wrote in his weekly report to Wakefield that, '... the old whaler, Barrett, died yesterday at Moturoa ...'[23] Barrett was aged forty.

They buried him at Moturoa's Waitapu urupa two days later, with 'a numerous body' of the original settlers at his graveside. Almost certainly, Rawinia, Caroline and Sarah would have had a strong group of Ati Awa supporters with them on the day too, but these tribal mourners' presence was not recorded by the English writers. So there Dicky finally lay, just up from the beach and at the foot of the slope where he had helped to defend Otaka pa. His monument was his cannon, half buried and spiked in the sands.

Governor Grey arrived in New Plymouth the day after Dicky's funeral, in response to the uproar FitzRoy's decisions had provoked. In a special

*Caroline (left) and Sarah Barrett. Sarah married William Honeyfield in 1852, Caroline married his brother, James, in 1864.* DAGUERREOTYPE 1852/3/TARANAKI MUSEUM

dispensation, he gave the grieving Rawinia and the two girls a lifetime interest in their family sections; then he, in his turn, put aside FitzRoy's awards, arranged for the settlers to have access to more extensive acres, and started trying to negotiate with Te Ati Awa again for the colonists to have the same amount of land in New Plymouth as Spain had awarded to the New Zealand Company.

Although Barrett's original purchase continued to be a subject of contention, over the ensuing decades his possible sins were forgotten. The second generation of settlers spoke his name with some affection, and perpetuated it in various road names and landmarks as New Plymouth developed: Barrett Lagoon, Barrett Reef, Barrett Domain, Barrett Road, Barrett Street and Barrett Street Hospital.[24]

\* \* \*

Putting together all that is now known about Dicky Barrett, a picture starts to emerge. He arrived in New Zealand a short, brown-haired, stocky young seaman, the sort you can imagine who rubbed his hands together and smiled with his pipe clenched between his teeth, and made people feel they were his friends. In the early years of partnership with Te Puni, Te Wharepouri and Te Ati Awa, he would have become well-muscled like them, living mainly on the same tribal fare. (The magnificent physiques of Maori men in those days were commented on by many visitors to New Zealand.) And the few references available suggest he was venturesome, energetic, full of initiative, adaptable, loyal and courageous.

In the six years he was at Te Awaiti Barrett became outstandingly corpulent. After earlier years on tribal rations, there was much easier access to European stores, and Dicky was no longer an energetic commercial traveller. He did not go out in the whaling boats; he had a more sedentary occupation as headsman and storekeeper on the station. Here he displayed kindness, generosity and friendliness, qualities shared, and perhaps fostered, by Rawinia. He embedded these traits into his management, and he and Love were accorded respect and liking by their employees. Several of them were associated with Barrett on and off throughout all his years in New Zealand, while the 'floating' whalers did not desert Te Awaiti in the same numbers as they did other stations. With this in mind, it would be easy to credit both Barrett and Love for together building up a well-known whaling establishment from practically nothing; but, in the light of Barrett's later business inadequacies, it must be presumed that Love was the main contributor to their success.

From 1839 to 1843, Barrett showed obvious ambition. He strove to rise above his previous station in life; but, under the weight of Colonel Wakefield's blatant disregard of his counsel, and despite the protestations of the chiefs from the other iwi, Barrett showed weakness after his elevation to the position of Agent for the Natives. A more forceful man would have pounded the table at vital advice being ignored. Barrett swallowed his disquiet and allowed his employer to carry him along. There was monetary reward for his involvement, as he ran with the hares and hunted with the hounds. While exploring for the Plymouth Company settlement site with Carrington he showed possible signs of guile, and used his local knowledge to influence the surveyor to site New Plymouth near Rawinia's birthplace at Ngamotu. He was criticized and blamed for it; but, comparing Nelson as a port with New Plymouth 150 years later, who is to say that Barrett was not being sincere in his guidance.

The only known sketch purported to be of Barrett,[4] evidently done after 1842, showed a man of average figure, perhaps the result of long periods of food shortages in New Plymouth. In the last years of his life, Barrett's positive characteristics were outweighed by his inability to avoid debt and his dismal failure in the whaling venture. But he was in good company here among many of the New Plymouth settlers; and, despite his misfortunes, people continued to mention his good-humour and entertaining and harmless tall stories, which equalled those of most of his peers. Like them, he may have had an 'eye for the girls', as one descendant reported hearing her mother describe him. Another strongly contradicted it. He was a real family man, she had heard. In those days, the one did not preclude the other. Most chiefs had children by several women, and Barrett spent a large proportion of his life in New Zealand among Te Ati Awa as an honorary chief.

This then is the legendary Dicky Barrett, one of only about 300 Europeans, nearly all men, living in New Zealand when his story began. Among that number were the missionaries, well-known historical figures who left copious written material for national archives and libraries. Other adventurers like Barrett, amongst those 300, would be remembered only in local districts around the country. But nearly every inhabitant of New Zealand knows of Barrett's Reef, and many have heard of Barrett's Hotel in Wellington, although Dicky was mine host for only about one year. Somehow, from his two scrappy letters and other people's journals, Barrett's singularity has become legend, and he remains one of New Plymouth's best known treasures.

# GLOSSARY

## General

| | |
|---|---|
| ahi ka | occupation rights, keeping fires burning on tribal land to retain ownership |
| ao-kai | daily food |
| ariki | first born in notable family/ spiritual or temporal leader |
| aruhe | dried, pounded fern roots |
| hangi | food cooked in umu |
| hapu | sub-tribe |
| harakeke | flax |
| heke | migrating party |
| hua | vegetables from ground, egg, fruit |
| hue | gourd |
| iwi | tribe |
| kai moana | seafood |
| kainga | village |
| kaitaka | fine cloak |
| kanga | curse |
| ko | digging stick |
| kareaeo | supplejack |
| kaukau | cow (from English) |
| kete | flax basket or kit bag |
| kokowai | red ochre clay |
| kopa-maori | Maori earth oven (adapted from copper) |
| korero | talk, discussion, |
| korerorero | long discussion |
| korito | cooked root of raupo |
| kotukutuku | New Zealand fuchsia |
| mana whenua | trusteeship of the land |
| maori | usual |
| maro | flax kilt, apron, girdle |
| mihinare | Maori convert to Anglicanism |
| mokomokai | dried heads |
| muka | processed flax |
| Onepoto | one of the cannon used in the defence of Otaka pa |
| ope | small, fast-travelling fighting force |
| pakeha | white-skinned person |
| panui | formal declaration |
| parepare | defences, earth wall |
| pataka | storehouse |
| pikau | travelling backpacks |
| pu | gun |
| Pupoipoi | a small field gun on wheels, used in the defence of Otaka pa |
| putangitangi | paradise duck |
| Ruakoura | one of the cannon used in the defence of Otaka pa |
| rangatira | chief, noble |
| rimurimu | seaweed |
| raupo | bullrush |
| Tama-te-uaua | 1832-33 heke, comprising Te Ati Awa and other tribes |
| taiaha | long hardwood weapon |
| take | beginning, cause, reason, subject of discussion |
| tangata whenua | local people/of the land |
| taua | war party |
| taua-iti | small advance force |
| taurekareka | slave/scoundrel, standard term of abuse |
| tetere | long wooden trumpet |
| tika | correct |
| toki | adze, axe |
| torotoro | tough creeping plant |
| tuahine | female cousins |
| tuku whenua | leased land, temporary rights to use tribal land, revoked when original user moved away or died |
| turangawaewae | place to stand, land rights, legitimate home |
| umu | oven |
| urupa | burial ground |
| utu | reciprocity |
| wai piro | alcohol, stinking water |
| waka koreru | small wooden trough with noose placed in trees to snare birds |
| whanau | family |
| whanganui | harbour |
| whare kohanga | birth house |
| whata | platform erected on wooden pole or poles, for storing food and keeping rats away |

## Glossary

**Place names**

| | |
|---|---|
| Anaho | tribal village on Arapaoa (Island) where several whalers had 'summer houses' |
| Arapaoa/Arapawa | large island in Queen Charlotte Sound |
| Fairhaven | John Guard's name for his whaling station at Te Awaiti |
| Hakaroa | Akaroa |
| Haowhenua | place on Kapiti coast where an inconclusive battle was fought in 1834 between iwi and hapu who had migrated there and settled too closely |
| Heretaunga | Hutt River |
| Hoiere | Pelorus Sound |
| Hongihongi stream | falls into the sea close to northern base of Paritutu |
| Huatoki | stream on northern side of New Plymouth |
| Jackson's Boat Harbour | small bay on the exposed eastern coast of Te Wai Pounamu, near entrance to Kura te Au (Tory Channel), where boats could take shelter in an emergency |
| Kaiwharawhara | Ngati Tama pa on shore of Whanganui-a-Tara |
| Kakapo Bay | site of John Guard's whaling station in Port Underwood |
| Kauraripe | Cloudy Bay |
| Kawhia | ancestral home of Ngati Toa iwi |
| Kororareka | Russell (Bay of Islands) |
| Kumutoto | pa on the shores of Whanganui-a-Tara, home to Wi Tako and his father, Ngatata-i-te-rangi |
| Kura te au | (red current; actually, myriads of minute crustaceans on which whales feed) Tory Channel |
| Mana | Entry Island |
| Matiu | Somes Island |
| Moana Raukawa | Cook Strait |
| Motunui | place in Taranaki where Ati Awa and Waikato warriors battled in 1821 |
| Moturoa | alternative name for Ngamotu area |
| Mangungu | Wesleyan mission station at the Waihou River |
| Mataora | one of the outer Sugar Loaf Islands |
| Mikotahi | the most impregnable Sugar Loaf island |
| Motuotamatea | one of the outer Sugar Loaf Islands |
| Ngamotu | the islands/the Sugar Loaf (now Sugarloaf) Islands; the area around them, also known as Moturoa |
| Ngauranga | Te Wharepouri's pa at Whanganui-a-Tara, south of Pito-one |
| Onapapata | name of bay where Jimmy Jackson had his whaling station |
| Onepoto | one of the cannon used in the defence of Otaka pa |
| Otaikokako | Ngamotu beach |
| Otaka pa | stood on a slope near where New Plymouth Cool Store is now sited |
| Otakou | Otago |
| Patarutu | the storehouse for holding trading goods, built by Ati Awa |
| Paritutu | 150-metre high, land-locked rock cone, one of the Sugar Loaf Islands; the other main islands are Mataora, Mikotahi, Moturoa, Motumahanga, Pararaki and Whareumu |
| Pipitea | Te Ati Awa pa on shore of Whanganui-a-Tara. Home to Wairarapa and Moturoa |

| | | | |
|---|---|---|---|
| Parininihi | white cliffs, north of Waitara, New Plymouth | Te Rimurapa | Sinclair Head |
| Pito-one | Petone, Te Puni's pa at Whanganui-a-Tara | Te Tangihana a Kupe | Barrett's Reef |
| Poneke | Maori adaptation of Port Nicholson | Te Uruhi | place on Kapiti coast where Ngamotu hapu of Ati Awa settled at the end of Tama-te-uaua |
| Port Cooper | Lyttelton | | |
| Port Nicholson | Whanganui-a-Tara/ Poneke/Wellington | Te Wai Pounamu | the South Island of New Zealand |
| Pukeariki | Main Ati Awa pa, which stood on a hill, since levelled, near present railway station in New Plymouth | Tiakiwai | small kainga on shore of Whanganui-a-Tara, a resting place for people on the track to Porirua |
| | | Totaranui | the Marlborough Sounds |
| Pukeatua | Te Wharepouri's home at Ngauranga | Turakirae | western point before entry into Palliser Bay at southern end of Te Ika a Maui |
| Pukenamu | small hill east of Wanganui, site of battle during Tama-te-uaua | | |
| Pukerangiora | fortified pa on a high bluff about 12 km from the sea, on Waitara River | Wairau | fertile plain on river inland from Kauraripe, where confrontation over land ownership occurred in June 1843. |
| Rakiura | Stewart Island | | |
| Rangitoto | D'Urville Island | | |
| Raukawa Moana | Cook Strait | Waiwhetu | a kainga on the north-eastern shore of Whanganui-a-Tara |
| Raurimu | a minor kainga on shore of Te Whanganui-a-Tara | | |
| Rekohu | Moriori name for the Chatham Islands | Wanga Nui Atera | misspelling of Whanganui-a-Tara in official documents |
| Rimurapa | Te Rimurapa, Sinclair Head | Whakaahurangi | name of the ancient track on which Tama-te-uaua began its walk to Kapiti Coast |
| Taitapu | part of Blind Bay/ Tasman Bay | | |
| Te Aro | pa on prime land on shore of Whanganui-a-Tara, occupied by Taranaki people | Whakatu | part of Blind/Tasman Bay |
| | | Whangaingahau | pa where Tama-te-uaua was received with welcome food towards the end of the trek |
| Te Awaiti | The Little Streams, site of Barrett and Love's whaling station | | |
| Te Ika a Maui | the North Island of New Zealand | Whanganui | name of the river at Wanganui, Taranaki |
| Te Karamuramu | place near Wanganui where members of Tama-te-uaua built temporary defences | Whanganui-A-Tara | Wellington Harbour |
| | | Wharekauri | Maori name for the Chatham Islands |
| Te Moana-a-Raukawa | formerly Cook's Strait; now Raukawa Moana | Wickett | European name for Arapaoa (Arapawa) Island |

260

# Glossary

## Individuals and Tribes

| | |
|---|---|
| Hera | Sarah, Dicky and Rawinia Barrett's third daughter |
| Hone Tanerau | John Daniel Love, Jacky and Mereruru Love's firstborn son |
| Kararaina | Caroline, Dicky and Rawinia Barrett's firstborn daughter. Also the name of Henare Barrett's Ngai Tahu mother |
| Hakirau | Jacky Love |
| Iwikau | brother of Te Heuheu Tukino II |
| Kurukanga | One of three chiefs who 'sold' Wakefield Wanganui land |
| Mahau | Te Ati Awa chief at Whanganui-a-Tara |
| Mananui Te Heuheu Tukino II | See Te Heuheu |
| Mereana | Mary Ann, Dicky and Rawinia Barrett's second daughter |
| Mereruru te Hikanui | Jacky Love's wife |
| Moki Te Matakatea | See Te Matakatea |
| Muaupoko | iwi displaced by migration of northern tribes to Kapiti coast, earmarked for extinction when they earned Te Rauparaha's wrath |
| Ngai Tahu | major iwi in Te Wai Pounamu |
| Ngatata-i-te-rangi | influential chief in Ngati Te Whiti hapu of Te Ati Awa but junior to Te Puni, Te Matangi and Te Wharepouri. Father of Wi Tako |
| Ngatata Wiremu Tako | Wi Tako, son of Ngatata-i-te-Rangi |
| Ngati Haumia | hapu of Ngati Ruanui iwi |
| Ngati Ira | ancient inhabitants of Whanganui-a-Tara, displaced by the influx of tribes with the great southern migration |
| Ngati Kahungunu | another iwi displaced by the influx of tribes into the Kapiti/Whanganui-a-Tara areas |
| Ngati Maru | hapu of Te Ati Awa iwi |
| Ngati Mutunga | iwi to north of Te Ati Awa. Their approximate boundaries stretched from Urenui to Pukearuhe, on Taranaki coast |
| Ngati Rahiri | hapu of Te Ati Awa iwi |
| Ngati Raukawa | iwi near Kawhia, allied to Ngati Toa |
| Ngati Rauru | iwi to south of Ngati Ruanui. Their approximate boundaries stretched from Patea River to Kaiiwi stream |
| Ngati Ruanui | iwi to south of Taranaki iwi. Their approximate boundaries stretched from Otakeho to Patea River |
| Ngati Tama | iwi to north of Ngati Mutunga. Their approximate boundaries stretched from Pukearuhe to Mokau |
| Ngati Tawhirikura | hapu of Te Ati Awa iwi |
| Ngati Te Whiti | hapu of Te Ati Awa iwi |
| Ngati Tuwharetoa | iwi in Taupo area |
| Nohorua | Te Rauparaha's half-brother and father of Te Ua, Joseph Tom's wife |
| Poharama | Te Ati Awa chief who supported early settlers in New Plymouth |
| Pomare | principal chief of Ngati Mutunga who gifted land to Ati Awa and |

| | | | |
|---|---|---|---|
| Puakawa | Taranaki chiefs when he led his people's migration to Wharekauri Te Ati Awa chief; strong opponent of selling land to William Wakefield at Whanganui-a-Tara. Subsequently murdered and mutilated, shortly after arrival of first settlers | Te Popo Te Puni Te Rangihaeata Te Rangitake Te Rauparaha Te Ropiha Moturoa Te Wharepouri Te Wherowhero Tiki Parete Tohukakahi Tokorehu Topeora Wairarapa Waitohi Wanganui Wakaiwa Wiremu Kingi Wi Tako | connections to Ngati Mutunga brother of Mananui Te Heuheu Tukino II senior Ati Awa chief with connections to both Ngati Te Whiti and Ngati Tawhirikura hapu Ngati Toa chief, son of Te Rauparaha's sister, Waitohi Ati Awa leader connected with Ngati Mutunga and Ngati Kura. He took the name Wiremu Kingi after his baptism in 1840 principal chief of Ngati Toa important Ati Awa chief, brother of Wairarapa leading Te Ati Awa fighting chief of Ngati Te Whiti hapu Waikato chief who belonged to the senior chiefly line of Ngati Mahuta Te Ati Awa name for Dicky Barrett Te Wharepouri's brother Tainui chief seeking further vengeance against Ati Awa after Pukerangiora sister of Te Rangihaeata Brother of Moturoa and uncle of Wi Tako Te Rauparaha's sister iwi to south of Nga Rauru; identified by the Whanganui River. Rawinia or Rangi, Dicky Barrett's wife *See both* Te Matakatea *and* Te Rangitake *See* Ngatata Wiremu Tako |
| Roka | Rose. Te Puni's daughter, just an infant during Tama-te-uaua | | |
| Tainui | confederation of tribes, including Waikato and Ngati Maniapoto | | |
| Tamaiharanui | leader of Ngai Tahu in the northern part of Te Wai Pounamu in the early nineteenth century | | |
| Taranaki | iwi to south of Te Ati Awa. Their approximate boundaries stretched from Omata to Otakehu | | |
| Tautara | principal chief of Te Ati Awa | | |
| Te Heuheu Tukino II, Mananui | famous principal chief of Ngati Tuwharetoa | | |
| Te Kaeaea | Ngati Tama chief, also known as Taringakuri | | |
| Te Manihera Te Toru | son of Te Matangi | | |
| Te Matakatea, Moki | principal chief of Ngati Haumiti hapu of Taranaki iwi. He took the name Wiremu Kingi when he was baptised in 1841 | | |
| Te Matangi | senior Ati Awa chief who was gifted land around Pito-one at Whanganui-a-Tara by Pomare when Ngati Mutunga migrated to Wharekauri | | |
| Te Peehi Kupe | Ngati Toa chief with | | |

# SOURCES AND NOTES

## Abbreviations

| | |
|---|---|
| AJHR | *Appendices to the Journals of the House of Representatives, New Zealand* |
| APL | Auckland Public Library |
| ATL | Alexander Turnbull Library, Wellington |
| BPP | *Irish University Press Series of British Parliamentary Papers, Colonies, New Zealand* |
| CP | Christchurch Press |
| DNZB | *The Dictionary of New Zealand Biography* |
| EGW | Edward Gibbon Wakefield |
| EJW | Edward Jerningham Wakefield |
| EP | *Evening Post* |
| FAC | Frederic Alonzo Carrington |
| H & N | D. Hamer and R. Nicholls |
| IHN | I. H. Nicholson |
| Jnl | Journal |
| JPS | *Journal of the Polynesian Society* |
| LFS | Letters from Settlers |
| NZH | *New Zealand Herald* |
| NZJ | *New Zealand Journal* |
| NA | National Archives, Wellington |
| NZC | New Zealand Company files in National Archives |
| NZG | *The New Zealand Gazette* |
| NZG & BS | *The New Zealand Gazette & Britannia Spectator* |
| NZG & WS | *The New Zealand Gazette & Wellington Spectator* |
| OHNZ | *Oxford History of New Zealand* |
| OLC | Old Land Claim (National Archives, Wellington) |
| RAHSJ | *Royal Australian Historical Society Jnl,* v. 2, 304–305 |
| R & S | Rutherford and Skinner |
| RB | Richard 'Dicky' Barrett/Tiki Parete |
| TH | *Taranaki Herald* |
| TM | Taranaki Museum |
| WW | William Wakefield |

## Prologue

Bloomfield, 1–212; RB's letter to brother, MS Papers 1183, ATL; WW's marriage cerificate, ATL, MS Papers 5745; Boulton, Introduction, 1; Chadwick, in *NZH*, 4 Jan. 1995, section 1, 8; *DNZB*, v. 1, 572–73; Evison, 36; Adams, 19–125; Burns 1989, 13–14, 25–26, 43, 80, 84; FitzRoy 1846, Chapter 3, 29; Harrop, 9–130; Johnson, 32; McDonnell, 1; McLean 1994, 6; O'Connor, 17–163.

1. At the same time that Gibbon was eloping with Ellen Turner, William was eloping to Paris with his wife-to-be, Emily Sidney.
2. Thousands of unemployed people had already emigrated to America, where each man became master of his own acres but could, of course, only bring into production as much land as he could till and sow and reap himself. Under Gibbon Wakefield's scheme, the labourer would take a few years to save enough money to buy land for himself, during which time his work would have contributed to the development of the new country. (See also Burns 1989, 25–28.)
3. Burns 1989, 14.
4. Ron McLean (183–84) found a Richard Latter Barrett's baptism (son of Matthew and Sarah Barrett) mentioned in the Mormon Records. He may be the Richard Barrett of this narrative, but the details cannot be totally authenticated and Barrett never included 'L.' or 'Latter' in his known signature.
5. Chadwick, *NZH*, 4 Jan. 1995, Section 1, 8.

## Chapter 1

Barrett 1962, 2; RB, Jnl, 1, 2, 3; Taranaki Education Board, 2; *DNZB*, v. 1, 481–83, 521–22; Evison, 54; McNab, 4–16, 18, 22–37, 41, 381–413; McLean 1994, 1, 28, 37, 41, 45, 184; Maning 44–58, 73–80, 112–60, 216–25; Marshall, 199-233; Mooney, 1; Morley, 79; IHN, 32, 33, 39; Parsons, July 1979, 3, 40; Parsons 1983, 56; Pevsner, 49; P.J. Power; Prickett, 7; Quin, 33, 34, 42; Scanlan 1985, 22; Seffern, 6, 13; Smith, 444–45; *Sydney Gazette*, 16 May 1829; WW, Diary, Notes 8; Wards, 216–17; Watene, 11.

1. Through the centuries, war had become stylized. It involved much preparation and travelling, much posturing, dancing and haranguing, followed by immediate hand-to-hand conflict or long-drawn-out sieges. The hostilities ended with final flight or by mutual consent. It was usually eminently satisfying for all participants. (Travers, 21–26; Maning 44–58) After the deaths of one, two or more chiefs in previous fights had been avenged, the attacking tribe would leave their foes and return home to harvest their crops. Whoever had been defeated accepted that the conquerors might take their crops or provisions; or take slaves from amongst their numbers, who would work for them, be wives for them, or be eaten as needs arose. But, unless peace was made at the time by both sides, they parted recognising that a day of further reckoning would be inevitable. With much inter-marriage through the generations, kin involvement became more widespread, and with the coming of the musket, casualties increased and the tribal balance was threatened.
2. All these Europeans died as a result of cultural misunderstandings. The explorers had infringed tribal taboos and been killed for utu.
3. After sailing in the South Seas for the previous six years (RB, Journal,1) or, possibly, while mixing with Maori in Sydney.
4. Smith, 445.
5. Hyndes and Street had built the *Adventure* in 1827, in a joint venture at Hyndes' timberyard in Cockle Bay, Sydney, and had sent her initially on Australian coastal trading trips. *Adventure* was a wooden, two-masted schooner, clinker built, 40' 8¼" long, 12' 3" wide, and 5' 9" deep, with one deck and a square stern (Parsons 1979, 40). Traditionally, trading ships came across from Sydney and sailed from place to place until they replaced the merchandise in their holds with the required goods, then they returned to Australia.
6. Ingram (1977, 7) says that Barrett and Love established a trading and whaling station on the Taranaki coast for Mr Thomas Hyndes, but the whaling appears to have started later. Ingram probably based his comments on others' writings.
7. Early travellers to New Zealand were impressed by the extensive, healthy, weed-free cultivations they saw around every Maori pa, industriously and constantly tended with simple stick implements. The Maori did not rotate their crops, but neither did they impoverish the soil. They simply left fallow the previously used plots, for fern and bracken to take over, cleared and burned another patch of bush and planted a fresh lot of seeds; but the Europeans could not accept this method. *See also* Beaglehole (Ed.) 'The Voyage of the Endeavour', 583–84, as seen in 'The Polynesian Foundation' by Janet M. Davidson in *OHNZ*, 21.
8. Her sisters were Harata Wai Kauri and Harata Pikia, neither of whom had children. Wai Kauri married Hoera Parepare (Taranaki Maori Board 39/70, New Plymouth; Judge Robert Ward, 22 Sept. 1898, Case 179 — Succession Ratapihipihi A. East: Deceased Harata Wai Kauri [*sic*]).
9. The Maori name came down through European writers, therefore there is a possibility her name was either *Tahora*, which means 'gather' (since there is uncertainty as to whether this group of seamen acted as whalers during this time) or *Toroa*, the Maori word for wandering albatross or mollymawk.
10. The ancestors of the dogs, like those of their owners, had come from Hawaiki many generations earlier in *Tokomaru* canoe.
11. Although barter was not tika — Maori commerce involved gift exchange — it was adopted of necessity when transactions involved Pakeha traders.
12. Some of the best *Phormium tenax* in New Zealand grew on the Taranaki coastline. Obtaining the fibre was a long and laborious job. Each leaf was first split down the middle, stripped of its outer edges with the nails of the forefinger and thumb, the bottom of it cut off, and a slight incision made across its dull side, either halfway or about 15 cm from the top. The firmly

grasped leaf was then drawn down lengthways over a sharp mussel shell, held in the other hand, thus stripping off its vegetable matter; the reverse side was stripped in the same manner. Each leaf yielded only 12–15 g of fibre. The fibres were then usually tied in small bundles and put into a running stream for the night; the following day, after being well shaken, they were hung to dry on a covered scaffolding. After this, the husk came off easily. The treated flax was never exposed to the sun or rain: the sun made it too brittle and the rain would discolour it. After three or four days' drying, it was fit for use, and could be twisted or woven into any article required (EJW, 44–45; Burns 1980, 131, 139).

13. The movements of the *Adventure* can be plotted fairly accurately. It is known to have left New Zealand on April 10 and arrived in Sydney May 4/6; and left there on the return voyage, via the Bay of Islands, on May 26/29 (IHN, 32). Given that the average sailing time between New Zealand and New South Wales was about twenty days, with the shortest ever recorded being nine days (IHN, 81, *Currency Lass*, in 1832), and the longest thirty-four days (IHN, 255, *Hannah*, in 1839), it is likely she was lost between June 21 and 26.

14. Te Rauparaha and Te Peehi were on a social visit to Tamaiharanui, hoping to obtain greenstone; but en route they stopped and killed many Ngai Tahu on the Kaikoura coast in revenge for a curse unwisely laid on Te Rauparaha by another Ngai Tahu chief (Burns 1980, 147–50).

15. It was 32–96 km wide and stretched 560 km up the middle of the island, all the way from Whanganui-a-Tara in the south to Manukau Heads in the north.

16. Te Ati Awa had joined a Waikato force and defeated Taranaki in 1826 at the battle of Kikiwhenua Okomakauru (Smith, 415).

17. Said to have been established by William Worcester, in the employ of John MacLaren and Co. in Sydney, who was joined by David Scott between 1831 and 1834 (Barrett 1962, 2).

18. The name is equivalent to Lavinia. Possibly adopted when Wakaiwa became a Christian; possibly chosen by Tiki in memory of some close female relative in England, although not mentioned in Mormon records of possible Barrett family members.

19. Elsdon Best, 'Maori Customs Pertaining to Birth and Baptism', *JPS*, v. 38 No. 152 (1929), 241–69.

20. *TH*, 21 Nov. 1899, 1.

21. Elsdon Best, 'Maori Customs Pertaining to Birth and Baptism', *JPS*, v. 38 No. 152 (1929), 241–69.

22. Called *Ameriki Watiti* by the tribes: Ameriki = Admiral, Watiti = Walker, the captain's name.

23. A different ship from the one that rescued the crew of the *Adventure*.

24. McNab 1913, 381–413; Burns 1980, 158–60.

25. Barrett's involvement is conjecture. Street's is recorded, since he had chartered the boat, and his less-than-honest behaviour is inferred from semi-official reports that 'strong pressure was being exerted by influential men in Sydney to get the accused and witnesses out of the road' (McNab 1913, 402–13). Captain Stewart was arrested, released on bail and disappeared for ever.

## Chapter 2

Bagnall, 5; Ballara, 13, 22, in H & N; Burns 1980, 122, 139, 142, 157, 163, 184, 189–91; Carkeek, 1; *DNZB*, v. 1, 315–16, 526; Evison, 50, 68–69, 77–78 Note 58, 88, 106; Grady 1978, 28–29, 48; Heaphy, 17; Heberley, Jnl, 24; Hood, 81; MacDonald, 62; McLean, Papers, Aubrey to McLean, 19 Dec. 1846; McLean 1994, 49–65; McNab, 18, 38–62, 70, 119; Maning, 42–44; *NZH*, 23 April 1892, Jervis article, Suppl. 1, col. 2, and 30 Nov. 1892, 1, col. 6; IHN, 59–81; Parsons 1979, 78, 1983, 56; Seffern, 6–13, 57; Smith, 359–75, 416, 460–79, 485–98, 538–40; Taranaki Report, 22; Taylor, 366, 383–84; Thompson, 140–41; EJW, 90, 228; Wards, 217.

1. Otaka pa was situated where the cool store now stands.

2. This pa was built 8 or 10 km from the sea on a 100 metre-high almost sheer

bluff beside the Waitara River.
3. It is said that Te Wherowhero personally despatched about 130 captives with the finest tattoos (Smith, 465). Twenty-seven years later, as a sick old man, he was elected the first Maori king. For a fuller account of the siege of Pukerangiora read Smith 1910, 462–67.
4. Tokorehu, of Ngati Maniapoto, was the still-aggressive chief. In 1821–22 he had been besieged by Te Ati Awa at Ngapuketurua when some sympathetic Ati Awa had created a diversion, enabling him and his followers to escape to safety into, ironically, Pukerangiora pa (Smith, 359–63).
5. Smith, 470.
6. Wi Tako survived to appear as an astute witness against the English eight years later. Much later, his daughter by his second or third wife, Josephine Ngatata Wi Tako, married Jacky Love's grandson, Daniel Kiri Love; and, in 1993, Dicky Barrett's great-great-grandson, Graeme Fairclough, married Jacky Love's great-great-granddaughter, Philippa Biss.
7. General Strange, while visiting New Plymouth on 15 July 1889, identified one of these, on display in Pukekura Park, as a '200-year-old valuable curiosity' (*TH*, 29 March 1941, 8, col. 2).
8. In 1833, eye-witness reports of the ensuing battle, written by Sheridan, appeared in the *Sydney Monitor* (McNab, 38). His account, and other word-of-mouth exaggerations and imaginative additions, contributed half the historical story. The other half came from J. S. Polack, a trader who lived in New Zealand 1831–37. He built up a picture of the battle from Ati Awa memories (Polack, 304–12). Unfortunately, little notice was taken of a series of articles written by H. M. Jervis which appeared in the *NZH* in 1892. He had been a storekeeper in New Plymouth in 1841 and had heard about the siege from Barrett's point of view, but then moved to Auckland, and interviewed Te Wherowhero in 1845, and obtained 'a Tainui view of the siege' (McLean 1994, 51).
9. Jervis, *NZH*, 23 April 1892.
10. Ibid.
11. Ibid.
12. Tohukakahi (Smith, 473 note 9, 477–78; Seffern, 7) was the father of the prophet Te Whiti o Rongomai III.
13. McNab, 47; Seffern, 8–9.
14. Smith, 473–74
15. Ibid; Jervis, *NZH*, 23 April 1892; Seffern, 7; McNab, 48.
16. Smith, 476; Seffern, 8; McNab, 52.
17. Sheridan, as quoted in McNab, 49.
18. Ibid; Smith, 474; Seffern, 8.
19. Jervis, *NZH*, 30 April 1892.
20. Or Buckell. Both names appear alternately in *Currency Lass* logs (IHN, 59–81).
21. Seffern, 10.
22. Although one account (Smith, 476, note 11) has the cannon captured by the enemy and exploding in their hands (see also Seffern, 11; McNab, 54–55; Smith, 476).
23. Skinner, *TH*, 22 Nov. 1899, 1.
24. Jervis, *NZH*, 30 April 1892.
25. Smith, 478.
26. Seffern, 13.
27. RB, Jnl, p.7.
28. For different accounts of the battle of Otaka see McLean 1994, 49–65; Seffern, 11–13; Smith, 462–79; McNab. 38–59; Jervis, *NZH*, 23 April 1892, suppl. p.1 col. 2, and 30 Nov. 1892, 1, col. 6.
29. It is difficult to be sure who these seamen were, since different accounts give different names. Jervis mentions only Love (Hakirau), Barrett (Tiki Parete), Wright (Harakeke), Daniel Henry Sheridan (Tameriri in Smith, 472; Tamirere from McNab, 41), Phillips, Akers and Oliver, an agent sent to Ngamotu by Montefiore of Sydney; but William Keenan (Hari Pataraki) was definitely there, and George Ashdown, Bosworth, Billy Bundy (Piri) and 'Black' Lee, from the crew of the *Adventure*, would have been in the neigbourhood.
30. Smith, 487–88.
31. Ibid, 488. Junction Road, New Plymouth, follows part of the track.
32. Ibid, 489.
33. Ibid, 490–92; Grace, 350.
34. Ibid, 492–94; Grace, 349–52.
35. The position is unclear. Kokohuia (Smith, 490 and 495) is mentioned later.

36. Smith, 493–95.
37. Ibid, 495 and note†.
38. Ibid, 496; Grace, 351–52, refers to this engagement as a Tuwharetoa victory.
39. Refer to whakapapa in Chapter 1.
40. Barrett told Dieffenbach in 1839, when he met him on the *Tory*, that they 'paid a courtesy visit to Te Rauparaha' (Ballara, 'Te Whanganui-a-Tara', in H & N, 22), which, taken with Smith's account of the heke's end, establishes the time of Ngai Tahu's attack on Guard's whaling station at Kakapo, and gives some indication of how Barrett and Love heard about these affairs and later came to take over Guard's station. It also indicates that Barrett had not been too fearful of Te Rauparaha, as Parsonson (1978, 196) suggests, to visit him with William Wakefield in 1839.
41. Carkeek, 1.
42. These had been created when Te Rauparaha took a Ngati Toa taua to the south of Te Ika a Maui, joining a marauding Ngapuhi taua, led by Patuone and Waka Nene. Te Rauparaha's private agenda had been to look for land to which he could remove Ngati Toa and also have contact with European trade and guns. He had 'found' the Kapiti coast, abounding in potential and apparently sparsely occupied, but had stirred up further strong antagonisms among the tangata whenua (Ngati Kahungunu, Muaupoko, Ngati Apa and Rangitane) when Ngati Toa eventually migrated there under his leadership.
43. Burns 1980, 122, 133–34, 139, 142, 157; Evison, 50.
44. Grady 1978, 48. This was common practice among chiefs of the time. They 'sold' land on the coast (or granted tuku whenua) to the local whale station owners, which gave the latter the right to fish for 12 km or so up and down the coast — the distance they could reasonably expect their crews to row a whaleboat.
45. McNab, 61–62. This wishful scheme never eventuated, but did alert people to the opportunities available in New Zealand, and led to increased trade between the two countries.
46. Smith, 498.

## Chapter 3

Adams, 74–81, 96, 169; Ballara, in H & N, 22–30; Barnicoat, Diary, 19 Feb. 1843; RB, Jnl, 1; Barrett 1864, 8, 21; Bigge, 701–718; *BPP*, v. 2, 617; Burns 1980, 177–179, 182, 193, 195, 237; Burns 1989, 79, 82–83, 85, 137, 175; Carkeek, 33–43; Crawford, 30–33, 40; Dieffenbach 1841, 13–17, 20, 29, 80–86, 116; Dieffenbach 1843, 28, 34–48, 49, 52–55, 62, 95–96, 116, 120; Dieffenbach 1844, Report to the New Zealand Company, 617; Darling 1829, 970, 975; Evison, 69, 70, 84–85, 86, 88, 90, 94, 96, 102 Note 16; Field, 63; Goddard, 19; Grady 1986, 45, 48, 164–65; Hayter, 23; Heberley 1996, 152; Heberley, Jnl, 27, 32a, 32b; *JPS*, v. 10, 99; Knocks, 1; Lethbridge, 50; McLean 1994, 2–3, 68–86; Morton 1977, 251–52, 254, 265, 1982, 32–44; IHN, 181; McNab, 1–3, 9–10, 19, 61, 70, 71, 135, 136, 138, 149, 151, 157, 196–97, 179–80, 188, 221–29, 230, 234–35, 265, 336, 433–59; Owens, in Rice, 32–33; Parsonson 1978, 177–78; 1995, 170 in Rice; HMS *Pelorus* log, Sept. 1838; Phillips, 33; *RAHSJ*, v. 2, 304–305; Smith, 497–500, 517–22; Seffern, 72; *TH*, 29 March 1941, 14; Thompson, 292–94; EJW, 24, 25, 32, 33, 36, 224–47; WW, Notes to Jnl. 18 Aug. 1839, 584, Index, 6, 8, 23, 59; Williment, 134, 136.

1. **The Sound.** Tory Channel was not then distinguished from Queen Charlotte Sound. Together they were both known simply as Totaranui.
2. In fact, shipping across the Tasman could not cope with the demands associated with whale products, and Barrett and Love actually inherited 240 tuns of Guard's stored oil (McNab, 61; Dieffenbach 1843, 55).
3. This name is to be found spelt both Toms and Thoms. Toms is the spelling on his gravestone at Te Awaiti, which gives his birthdate as 1799. One description has Toms as being 'of small stature and repulsive features' (EJW, 33). Another has him as a 'short, stout man with a trunk like a barrel and looking everyone straight in the face', with a 'lusty voice' and 'a good sort of fellow on the whole', despite his method of tying up and holding prisoner any employee who

disputed his orders (Crawford, 32–33).
4. Known as 'Oraumoa-nui' (Evison, 69–70).
5. Burns 1980, 178–79, 189.
6. Heberley, Jnl, 27.
7. Tall columns with pulleys that hauled the blubbery flesh from the carcase and kept it hanging before it was sliced up and processed.
8. Morton 1977, 251–52.
9. There were eventually close to 1,000 Te Ati Awa in the area.
10. Crawford, 40.
11. These had been built up in consultation with other masters (for examples, see McNab, 196–97, 447; EJW, 229).
12. *Balaena australis antarctica* or *Balaena antipodarum* (Dieffenbach 1843, 44), or *Balaena glacialis* (Orsman, 901) the black, or right (southern right) whale. The fish were generally black, stocky and fat, averaging 15 m in length and 55 tonnes in weight. (For further descriptions see Dieffenbach 1843, 41–48; whales, 52–55; extinction predicted, 48–52.) The enormous leviathans arrived off the coast of New Zealand in early May to begin calving, skirted the western coastline of Te Ika a Maui (the North Island), swam between Kapiti and the mainland, and down through Moana Raukawa (Cook Strait) to Kauraripe (Cloudy Bay). By June, they reached Wharekauri (Chatham Islands); then they turned back. By Sept., they would be found swimming east or north again and by early Oct. they were gone.
13. The *Mary Mitchell* was an American bay-whaling ship anchored in Cloudy Bay for the 1836 season. If her log is typical, for well over half the season boats were out in winds ranging from fresh to strong or gale force, and accompanying seas. Often, calm mornings could turn into afternoon gales in which the boats would have to run for home, sometimes leaving harpooned whales flagged. Out of the 149 days of that season, there were only seventeen on which the *Mary Mitchell*'s boats did not go out at all. (McNab, 433–59).
14. Grady 1986, 148; EJW, 230; Morton 1982, 245.
15. Morton 1982, 36.
16. EJW, 230.
17. Jervis, *NZH*, 23 April 1892.
18. Tamihana Te Rauparaha as seen in Burns 1980, 182. Although she puts this comment in 1836, it must have been 1835. There were too few whales caught in 1836 to produce this result.
19. Smith, 497–500.
20. *DNZB*, v. 1, 348.
21. Parsonson 1978, 177.
22. Ngai Tahu were cut down in numbers none of their enemies had ever imagined possible. It has been estimated that their numbers were halved (Evison, 86). (An old tohunga mourned: 'Before we were visited by ships, disease was rare amongst us. Few died young. Now few live to be old' [J. Watkin, MS 'Jnl', 5.7.1842, as seen in Evison, Note 16, 102]. Both races were defenceless before tuberculosis, but Maori died much more quickly.) It was possibly at this time that Love contracted the illness that killed him in 1839, although some Love family members of the present generation under-stood that he died of 'war wounds'.
23. Burns 1980, 193; McNab, 222-23. (McNab does not mention the wild turnips. He writes that the antagonism started when a canoe was smashed as the result of 'some dispute'.)
24. Heberley, Jnl, 32b.
25. McNab (223) gives the date of this battle as Feb. 7, and says that it ties in with Hempleman's log, where the 'Otago Maoris' were travelling north 'on fighting bent' 9–12 Jan., and returning home 12–14 Feb. (see Evison, 94; Parsonson, *OHNZ*, 170, 'Ngai Tahu Food Growing and Gathering'); yet Heberley (p. 32b) says that the fighting was 'about a man and another man's wife', and was going on when they reached Te Awaiti. The *Hannah*'s log gives their arrival there as Feb. 14 1838. Since the Otago Maoris were not involved in this conflict, Feb. 14 seems a more accurate date.
26. McNab, 222.
27. Burns 1980, 193.

## Chapter 4

Boulton, Intro. 1; Burns 1980, 179, 203, 237; 1989 100, 113; Carkeek, 64; Caygill, 79; Clarke jnr 1861; Crawford, 29; *DNZB*, v. 1, 348, 451; Dieffenbach 1843, 30–66, 144; 1841, v. 1, 31; *EP*, 15 Aug. 1929, 12, col. 3; Evison, 94–96; Field, 72; Grace, 141–43, 344–49, 351–52; Grady 1978, 29; *BPP*, v. 2, 66, 588, 590–96, 598; Ballara, in H & N, 20, 22–34; Harrop, 60–161; Heaphy, 3, 4; Heberley, Jnl, 33; Johnson, 32; Knocks; Lethbridge, 50; MacDonald, 65; McLean 1994, 95–97, 102, 106–108; McNab, 137–42, 347, 349, 350; NZC 3/1 WW to Directors; OLC 906 Box 45, Case 374, May 10–20 1842, evidence of Wi Tako, Mahau, Te Kaeaea and Te Puni, 8–18 Feb. 1843, RB's evidence; Parsonson 1978, 185–87; *RAHSJ*, v. 2, 304–305, v. 4, 134, 305; Smith, 533–34, 551–52; EJW, 14–60; WW, Notes to Jnl, 29 Aug. 1839 & ff; Wards, 216ff; Williment, 135.

1. Family legend has Love dying of war wounds suffered at the battle of Otaka pa or Kuititanga. The former was over seven years previously, the latter took place six weeks after the *Tory* dropped anchor at Te Awaiti, by which stage Love was already reported as dying. It is possible he was wounded during the attack by Ngati Toa on Te Awaiti, in early 1838, when he had just brought his family back from Sydney in the *Hannah*, or that he had contracted tuberculosis.
2. Dieffenbach 1843, 36.
3. OLC 906, Box 45, Case 374, 10 May 1842, WW to Clarke; NZC 3/1 WW to Directors, 77.
4. EJW, 28; NZC 3/1, 31 Aug. 1839, WW to Directors. It is unclear whether Barrett was offered a monetary inducement. Two or three years later, in a letter to his brother (see Bibliography [Letters]), he wrote that he was offered 'a Considerable Sum', but that may have referred to the £100 he was paid by a grateful William Wakefield when the purchase deeds for New Zealand had been signed with Barrett's help.
5. OLC 906, Box 45, Case 374, 8 Feb. 1843, RB to Clarke.
6. EJW, 26.
7. *BPP*, v. 2, 593.
8. EJW, 14; Written 'Nayti' by EJW. Probably Ngaiti or Ngati. Ngaiti had travelled to France on board a ship that had anchored near his village in Cloudy Bay during the 1836 whaling season (McNab, 245). He had then become a ward of E.G. Wakefield's in England for two years, and been designated the New Zealand Company's expedition's interpreter, in return for his passage home; but W. Wakefield was already doubting his ability and credibility. Ngaiti had showed uncertainty in his dealings with the New Zealanders they had already met and had been ignored by them. It subsequently turned out he was not of chiefly rank and had quite overstated his tribal position in England.
9. The catch of whales dwindled from a peak of seventeen caught by one Te Awaiti station in 1833, to nine caught by the three stations in 1839 (Caygill, 79).
10. Burns 1989, 100.
11. Instructions, signed by John Ward, for the guidance of WW, Principal Agent for the NZC, after the arrival in New Zealand of the ships about to sail for the first colony; including Letter from New Zealand Land Company to Colonel Wakefield, 16 Sept. 1839, No. 8, MS Papers 704, New Zealand Collection Folder 1, NA.
12. EJW, 28.
13. Ibid.
14. OLC 906, Box 45, Case 374, 8 Feb. 1843, RB to Clarke.
15. EJW gave his version of this in *Adventure in New Zealand*, 27–28. EJW returned to England with a tarnished reputation, after purportedly sowing his young man's oats in profusion in New Zealand, but his memory for detail and humour make his book enjoyable reading, so long as it is balanced by being seen as an advertising vehicle for the NZC.
16. *BPP*, v. 2, 590.
17. Ibid.
18. EJW, 28.
19. *BPP*, v. 2, 590.
20. EJW, 37. Waiana to the locals (MacDonald, 65), reputed to be a London

solicitor (Field, 72), or a Dutchman and 'almost a gentleman' (Lethbridge, 50).
21. Heberley, Jnl, 33; EJW, 49.
22. *BPP*, v. 2, 594.
23. OLC 906, Box 45, Case 374, 8 Feb. 1843, RB to Evans.
24. *BPP*, v. 2, 594; EJW, 52–53.
25. For different interpretations of the situation, read Wards, 216–19; Grace, 141–43, 344–49, 351–52; Burns 1980, 179; Ballara, in H & N, 22–34; Smith, 533–34, 551–52; McNab, 137–42; EJW, 52–53; Parsonson 1978, 186–87.
26. OLC 906, Box 45, Case 374, 8 Feb. 1843, RB to Evans.
27. Burns 1989, 113.
28. See Note 11.
29. Ibid.
30. OLC 906, Box 45, Case 374, 8 Feb. 1843, RB to Clarke.
31. EJW, 57.
32. Ibid, 58.
33. *BPP*, v. 2, 596.
34. EJW, 59.
35. Dieffenbach, *New Zealand and Its Native Population*, 22, as seen in McLean (1994) 106, note 38.
36. EJW, 59–60.
37. Several of the chiefs present would have been to Port Jackson, been aware of how Aborigines had been treated, and put two and two together about their own future prospects if they alienated their land.
38. *BPP*, v. 2, 596.

## Chapter 5

Barrett 1962, 8; *BPP*, v. 2, 159, 596, 597, 641; Burns 1980, 204; Clarke jnr 1903; Dieffenbach 1841, 87; *EP*, 16 Aug. 1929, 3; McLean 1994, 103; McNab, 359; NZC 3/1, 100, 106, WW to Directors; OLC 906, Case 374 Box 45 10–20 May 1842, WW to Spain, Wi Tako to Spain, Te Puni to WW, Dorset's evidence, Te Wharepouri to Evans, following Te Kaeaea, RB's evidence 8–18 Feb. 1843; Parsonson 1978, 185, 191; Rusden, 193; Taylor, 233; EJW, 60–66, 68–71, 73 –75, 384; WW, Jnl, 97–101; Ward 1928, 302, 466; Williment, 132–33.

1. OLC 906, Box 45, Case 374, Feb. 1843, RB to Evans.
2. However, when Henry Williams had sent his message to Whanganui-a-Tara in Sept., Reihana had complicated the issues and lost his validity by reverting from Wesleyan back to the CMS.
3. EJW, 61.
4. OLC 906, Box 45, Case 374, Feb. 1843, RB to Evans.
5. Ibid.
6. Ibid.
7. *BPP*, v. 2, 597.
8. EJW, 65; OLC 906, Box 45, Case 374, 20 May 1842, Wi Tako to Spain.
9. See Chapter 4, 7.
10. EJW, 62.
11. Ibid, 66.
12. Copy of Deed 374, 27 Sept. 1839, OLC 906, 45/756 Enclosure No. 28.
13. OLC 906, Box 45, Case 374, Feb. 1843, RB to Evans.
14. Ibid, 13 Feb. 1843, RB to Clarke.
15. Clarke 1903, 48–49.
16. As J. C. Crawford, who arrived at Port Nicholson a few days later, testified: 'The women and children of tender age often give their opinion and are listened to with respect' (*BPP*, 18 June 1844, v. 2, 159).
17. WW, Jnl, 100–101.
18. OLC 906, Box 45, Case 374, Feb. 1843, RB's evidence; EJW, 68.
19. Taranaki Street to Kent and Cambridge tces. The present Basin Reserve was a swamp called Waitangi (Ballara, in H & N, 25).
20. The chapel had been built on the eastern side of Te Aro stream. The apex of the triangle of land earmarked by Bumby and Hobbs was the intersection of Manners and Dixon streets, with Cuba Street forming the base of the triangle (Williment, 200). Bumby described the area bought in Port Nicholson: 'Going by this side of the river Te Aro, to where the river breaks into the sea, by the seaside to the broken hills of the land, and turning upwards along the ridges and spurs, turning to a valley, swamp, and falling down into the river Te Aro . . .' (Morley, 79, col. 1).
21. Parsonson 1978, 185.
22. But what two-facedness by Te Wharepouri. Bumby and Hobbs landed at Port Nichol-son on 7 June 1839 and Te

Wharepouri accompanied them as they explored the harbour in their whaleboat. Although Hobbs was very fluent in the Maori language, Te Wharepouri, presumably, acted as facilitator for their verbal agreement to purchase provisionally the 'no man's land' between Te Aro and Kumutoto, followed by a deposit of goods as an 'earnest' intention to buy. On 13 June, they attended a feast to celebrate the purchase for the mission (Williment, 132–34).
23. Barrett 1962, 8.
24. EJW, 73.

## CHAPTER 6

Angus, 275; BPP, v. 2, 66, 622, 628–30, 632, 634; Burns 1989, 119, 203, 208; Carkeek, 55–56; Dieffenbach 1843, 94–133; Evison, 113–14; Heberley, Jnl, 33–35, 37; Jackson, 3; Knocks, 14; Lethbridge, 43, 51; McLean 1994, 113–14; McNab, 112–32, 236ff, 280, 290–92, 352, 354; NZC 3/1 202; OLC 910–14, Case 374D, WW to Evans 17 June 1842, RB's evidence 31 May 1844; NZJ 1840, v. 1, 113; Parsonson 1978, 193–94; Quin 5–6; Rusden, 195–97; Seffern, 28–29; Smith, 461 note, 497–500, 524–33; Taranaki Report — Kaupapa Tuatahi, 35; EJW, 78–80, 97–100, 102, 104-109, 203–11; WW, Diary, 62 Note 5, 101, 106–109, Jnl, 630, 634–35; Wells, 20; Williment, 132–38, 140.

1. Angus, 275; EJW, 99, says the canoe was painted red and black; WW noted it as black and white (BPP, v. 2, 622).
2. Evison, 115 and note 23, 122.
3. EJW, 80; McNab, 352.
4. WW had not followed his instructions to coddle and promote Ngaiti, whom he found 'worse than useless as an interpreter ...' (NZJ, Sunday 27 Oct. 1839); BPP, v. 2, 622; EJW, 92–97.
5. OLC 910, Box 1, Case 374 C, Deed 25 Oct. 1839; Evison, 114; Burns 1980, 206-8 and Appendix II (re Deed), 318–20.
6. Parsonson 1978, 193–94.
7. At an estimate, the number of acres between the 38th and 43rd degrees latitude in New Zealand is two million at the least. Given the value of the goods that WW had for his purchase, this works out at about a farthing an acre at the most.
8. Originally brought by RB and Love and other visiting seamen (BPP, v. 2, 629; Dieffenbach 1843, 119–20; EJW, 99).
9. Ngati Tama, Ngati Rakei, Ngati Mutunga, Taranaki, Ngati Apa, Ngati Kuia, Rangitane, among the better known.
10. BPP, v. 2, 629 and v. 12, 290, appendix, WW's first despatch extracts, Saturday 2 Nov. 1839; OLC 906, Case 374B, Queen Charlotte Sound Deed, 8 Nov. 1839; WW, Nov. 1839.
11. EJW, 100.
12. BPP, v. 2, 66; Heberley, Jnl, 33–34.
13. BPP, v. 2, 630.
14. WW did not report this visit to his directors (Burns 1989, 208).
15. WW, Notes to Jnl, 107–108.
16. Burns 1989, 208; EJW, 105.
17. BPP, v. 2, 632; Dieffenbach 1843, 128. This deed was not signed and sealed until May 1840, when EJW completed the deed for his uncle (EJW, 104, 203–11).
18. WW, Jnl, 22 Nov. 1839.
19. Identified as the grandson of the principal chief of all Te Ati Awa, which makes him a cousin of Rawinia (WW, Notes to Jnl, 62).
20. Heberley said it was three men who jumped into the sea from the inner Sugar Loaf island (Heberley, Jnl, 36).
21. HMS Alligator, sent to rescue the Guard family captured by tribespeople after being wrecked on the Taranaki coast (McNab, 116–19; Marshall, 149–233).
22. Evidence presented later to Commissioner William Spain when he was investigating the legitimacy of the land sales in Taranaki (OLC 910–914, Case 374D, 17 June 1842, WW to Evans; ibid, 31 May 1844, RB to WW).
23. Seffern (29) incorrectly says that Dorset was landed with them. Dorset went north with WW, and returned with EJW in the Guide.

## CHAPTER 7

RB, Jnl, 5; Biggs, 48–50; BPP, v. 5, 50, 52, Dorset to Spain; Burns 1989, 167; Dieffenbach 1843, 133–71, 195; Heberley, Jnl, 37, 39; Hursthouse 1849, 11, 52; Jackson, 38; McLean 1994, 116, 117 note 81, 118–120; McNab, 292–93; Mullon, 7, 11; OLC,

910, Case 374D & E, RB's evidence 31 May 1844, 12–37, Edward Pukaki evidence 3 June 1844; *NZH*, 23 March 1892, 1, cols 4–5; Quin, 33–34; Rusden, 206; R & S, 64; Scanlan 1985, 9; Seffern, 30; Skinner 1921, 6; *Taranaki News*, Whiteley Suppl., 24 Dec. 1899, 6, col. 5; Taranaki Report, 24–25 - 2.3.2, 33 -2.4.2, 35 -2.4.4; Thompson, 272; EJW, 114–18, 121, 123–27, 131–35, 151; Wells, 41–42 says 72 people; Whiteley, Letters, 1552 13 May 1841, 1–5; Skinner, 'The Gospel Comes to Taranaki', pamphlet.

1. White had been dismissed from the ministry for 'excessive commercial activity and misapplication of mission property', but his choleric temper and uncompromising personality may have contributed to his downfall. History concedes that some of his actions were intended to be, and were, beneficial to the Maori (Laurenson, 29–30; *DNZB*, v. 1, 589; Parsonson 1978, 231).
2. Seffern, 30; Jackson, 38; Dieffenbach 1843, 134–39.
3. EJW, 123.
4. OLC 910–914, Case 374 D, 16 June 1842, John Dorset examined by Evans.
5. OLC 910–914, Case 374 E, 17 June, WW to Evans.
6. Mullon, between 16–18. Puke was probably Rawinia's father.
7. WW, Jnl, 25 Nov. 1839–11 Jan. 1840.
8. EJW, 121. Coincidentally, the *Guide* was owned by Thomas Street at that time (Parsons 1983, v. 1, 84). Clarke jnr thought the *Guide* had come from Port Nicholson ('Remarks on Busby's Letter', Clarke jnr to snr).
9. Parsonson 1978, 231; Whiteley, Jnl, 1501–2; Whiteley, Letters, 13 May 1841, 1–5.
10. Taranaki Report, 33 2.4.2; EJW, 124. RB may possibly have been culpable of 'insider trading' over the buying of the land at Whanganui-a-Tara. He could be accused of irresponsibility over the inaccuracies in the deed at East Bay (although when is an interpreter deemed to be responsible for his employer's actions?); but no one could have expected him to be *au fait* with the complexities of the ownership of tribal land in Taranaki.
11. EJW, 126–27, 131; Biggs, 48–50.
12. Ngamotu Deed, OLC 910, Box 1, Case 374D.
13. Taranaki Report 33 2.4.2. and 35 2.4.4.
14. EJW, 133.
15. Ibid.
16. Until the early months of 1841, the control of New Zealand affairs was in the hands of the Queen's representative in New South Wales, Australia.

CHAPTER 8

Adams, 160, 252; Barnicoat, Jnl, 12 Feb. 1841; RB, Will, SEP 317; Boulton, Intro., 2; *BPP*, v. 2, 304, 630, 636; Buller, 77; Burns 1989, 14, 49, 73, 76, 104–105, 107, 120–21, 129–30, 133–34, 136 and note 9 on 314, 152, 155, 171–72; Carmen, 2–7; Chapman, 3 Feb. 1847, ATL; Crawford, 33–39, 47–48; *DNZB* v. 1, 248, 360, 522, 589; Dieffenbach 1841, v. 1, 19; Heaphy, 68; Lawn, 1–3; McNab, 343, 344, 357, 360–62; NZC 1/28, 31 March 1840, RB to WW; NZC 3/1 269, 277–78, 19 Jan. 1840, 303–305, 27 Feb. 1840, 265–66, 27 March 1840, 348, 31 March 1840, WW to Secretary; NZC 3/1, 319–44, 3 March 1840, Hanson to Ward; NZC 3/2, 165, 25 May 1840; NZC, 31 March 1840 & No. 37, 7 June 1840, Revans, Daniell and St Auburn & No. 43, July 1841, RB to WW; NZC, 1st Report, v. 1, 14 May 1840, 13, 15, 21ff; *NZJ*, 10 April 1841, 21 Aug. 18; IHN, 240; OLC 906, Box 45, Case 374, 20 May 1842, Wi Tako to Spain; Parsonson 1978, 25; Petre 1971, 12; Revans, 23, 26, 38, 43–44, 73 and 18 April 1840; Owens, 52, in Rice; Rogers 1973, 234–45; Taranaki Report, 23; Taylor, 38; EJW, 51, 114–21, 136, 142–45, 148–55, 199, 203–12; WW, Notes to Jnl, 109ff; Ward 1928, 23–30, 35, 39–40, 43, 46, 49–51, 61, 74, 120, 467; John Ward to WW re Instructions, Letter from New Zealand Land Co. to Colonel Wakefield, 16 Sept. 1839, 21. No. 8 MS Papers 704, NZ Collection, folder 1, ATL; Williment, 134–35, 200.

1. Built by Te Puni, who had taken responsibility for looking after the stores. Colonel Wakefield occupied one corner of it until he moved to Thorndon.

2. Over 600 Europeans had sailed in, of whom about 180 were women, only a dozen or so being single. About thirty-three women had come in the *Aurora*, seventy in the *Oriental*, forty-one in the *Duke of Roxburgh*, and thirty-six in the *Bengal Merchant*.
3. Te Puni and Te Wharepouri later dismissed Tod's right to these acres, saying that Moturoa was only an insignificant chief. In reality, Moturoa was a senior chief, who had made the land available to Tod to demonstrate that the colonel had no power over it. But the passage was typical of the power plays constantly used among the chiefs of the various tribes, which repeatedly confused the white people (McNab 343; Ward 1928, 39).
4. Ward 1928, 43.
5. McNab, 344.
6. EJW, 149.
7. The name of the country estate of the chairman of the New Zealand Company, George Lambton, now the Earl of Durham.
8. Revans, 18 April 1840.
9. *New Zealand Jnl*, 10 April 1840 and 21 Aug. 1842, as seen in Ward 1928, 57.
10. *BPP*, v. 2, 636, WW to NZC Secretary, 27 Feb. 1840; NZC 3/1, 277–78; Burns 1989, 136.
11. By comparison, one William Gilbert, a sawyer already well ensconced, was earning £3 a week, and one watchmaker earned as much in a day. Captain Chaffers was later appointed harbour-master at £300 pa.
12. Burns 1989, 97; NZC 3/1, 372–80, WW, despatch of 25 May 1840.
13. John Ward's letter to WW re Instructions.
14. Burns 1989, 152; Ward 1928, 49.
15. Ward 1928, 51; Dieffenbach 1841, v. 1, 19.
16. *DNZB*, v. 1, 360.
17. Revans, 73.
18. The mission house and chapel, which the mihinare had built, stood out in the middle of the no-man's-land between Te Aro and Pipitea. The chiefs at Te Aro and Pipitea had actually promised a precisely defined area to Bumby and Hobbs, which they had left Reihana looking after, along with other native teachers. But Reihana reverted from Wesleyan to CMS after they left, and began speaking about 60 acres that he had reserved for himself. By the time that Henry Williams came to Whanganui-a-Tara, just after the *Tory* left on 4 Oct. 1839, Reihana wanted to return to his home in Taranaki, and Williams wrote that he had bought the land from Reihana to hold for the CMS. Barrett would not have known this.
19. This letter is to be found in the National Archives (NZC 108, 31 March 1840, RB to WW). It appears to be a copy, since the signature is not the same as that which appears at the end of RB's original Will, and is spelt with one 't'. It was usual to have official documents copied, but, in this case, the original should be there too. It can be wondered whether WW may not have had the letter 'arranged', so as to have an apparently genuine document, should his rights to that land be questioned.
20. And yet there was no great stigma attached to children from liaisons outside marriage. Adultery was punished, but not the fruits of it. In times of war, women were constantly being 'made wives' of victorious chiefs; and the tribal rights of the children from those unions were acknowledged. Tiki Parete had status in Rawinia's hapu and iwi, and it would not have been considered too unusual for him, or any chief, to have liaisons with women other than his wife. And girl babies were killed, but not boys.
21. And Octavius Hadfield had established a station at Waikanae at about this time. This account came from a whanau member connected with the Barrett Family Hui held in Christchurch over Labour weekend 1996. There are other Barretts who claim descent from Barrett, but have been proved to have descended from Nathaniel Barrett, a Wesleyan teacher from Kawhia, working there about 1844-47? onwards. None of this, unfortunately, is able to be validated. One Barrett descendant in the 1990s commented that her grandmother had said

that Barrett 'always had an eye for the girls'. Another, with just as much emphasis, says that his grandmother said that Barrett and Rawinia remained a close and loyal couple. DNA testing would not be conclusive. In the timetable of Barrett's life March–April 1840, there are no day by day records.

22. McNab, 362; Burns 1989, 155.

## CHAPTER 9

RB, Jnl, 2; RB, MS Papers 1183, ATL; Barrett 1962, 2, 8; *BPP*, v. 2, 61–62, 6 June 1844; *BPP*, v. 3, 84, Shortland to WW; Burns 1989 59, 139, 147, 152–54, 156–57; Bremner, 151–63, in H & N; Carmen, 10–12; FAC, Letters (in Jnl APL & Papers TM), 12, 16 Dec. 1840, 22 Sept. 1841; *CP*, 16 July 1982; Crawford, 33–39, 45–46; *DNZB*, v. 1, 400, 593; Easdale, 14–22; Field, 11; Harrop, 71; Heaphy, 8–16; Johnson, 37–39; Lawn, 1ff; McLean 1994, 127–29, 131, 132 note 23, 135–36; McNab, 359, 365; Moore, 57, 88–89; Natusch, 110; Natusch and Swainson, 128; NZC 3/1, 26 Jan. 1840, 305–307, 375–78; NZC 108 2/2, 4 Sept. 1841, Hanson to Halswell, as cited in McLean 1994, 134; NZC Reports, 3rd Report, 1 May 1841, 14, WW to Dillon Bell; NZC Reports, v. 1., 16; *NZG*, 25 April, 16 May, 6, 13, 27 June; *NZG & BS*, 29 Aug., 24 Oct. 1840, *NZG & WS*, 19 Dec. 1840; *NZJ* 1850, 293, W. Partridge to Chapman; OLC 906, Box 45, 13–16 Feb. 1843, RB to Clarke; OLC 906, Box 45, Deed signed 29 Aug. 1840; Parsonson 1991, 28, 1995, in Rice, 167–76 ; Petre 1971, 23–24, 31; Revans, 41, 52–53, 60, 68; Rogers 1973, 234–35; Seffern, 36–38; Taranaki Report, 26 2.3.3; Taylor, 48–49, 51, 231–32, 238–39, 245, 247–48, 260; Tonk, in H & N, 45–46; Turton, 162–63; EJW, 196, 199–203, 221, 251–55, 258–62, 281–85, 297–98, 301; WW, Notes to Jnl, 23, 40; Ward 1928, 30, 50–52, 56–57, 59–61, 66–71, 403, 487.

1. *NZG*, 25 April 1840.
2. EJW, 203.
3. *NZJ* 1850, 203, W. Partridge to H. S. Chapman.
4. Moore, 88–89.
5. Moore, 57, says that he bought it; Johnson, 37–39, has Revans as owner; Ward 1928, 467, has the *Jewess* as a 57-tonne schooner.
6. McLean 1994, 129–30, writes that RB indicated that the blankets were presents, but his source (OLC 906, Box 45, Case 374, RB evidence, 8 Feb 1843) could probably refer to a later date when RB and Brookes left blankets on the beach at Te Aro.
7. For some details see Rogers, 235–37. For further details, ibid notes 16 and 17.
8. Wairarapa and Moturoa were Wi Tako's 'uncles'.
9. See Chapter 8, 143–44.
10. WW, Jnl, 25 May 1840.
11. *NZG*, 30 June, 27 June 1840.
12. Burns 1989, 156; EJW, 249–51.
13. This comprised 1,000 sections for the thousand holders of land-orders, and 100 sections to be native reserves, the chiefs having been promised one section in ten. It will be seen that by now the ratio was actually one section in eleven.
14. Burns 1989, 157; Ward 1928, 59–61.
15. An important resting place for all iwi on the road to Porirua.
16. '. . . slight differences', occurring occasionally, is how Henry Petre (31) glosses over these in his booklet about the colony.
17. Although, later, they said they were to allow the surveyors' pegs to be 'unmolested' (Taylor, 51).
18. *BPP*, v. 3, 120, Shortland to Hobson 9 Oct. 1840.
19. OLC 906, Box 45, Deed signed 29 Aug. 1840, 'Agreement with the Natives of Pah Taranake'.
20. RB, MS Papers-1183, ATL.
21. Governor Gipps was still senior to Hobson. The Legislative Council was his law-making arm and the official channel for passing land-purchase legislation in New Zealand.
22. Among them, a mathematically intricate formula awarded one acre of land for every sixpence worth of goods given between 1815 and 1824, and one acre for every four to six shillings' worth of goods up to 1839; but the goods were to be valued at three times the Sydney price.
23. Revans, 52–53.

# Sources & Notes

24. Dr Evans, Henry St Hill and Henry Moreing were the delegates sent to see Gipps in Sydney.
25. See note 17 Chapter 1.
26. Taylor, 248
27. *NZGBS*, 24 Oct. 1840, 3; Ward 1928, 70–71.
28. Taylor, 260.

## Chapter 10

Aubrey, Jnl, in McLean 1994, 137–57; RB, Will; *BPP*, v. 2, 62–63, 69; Burns 1989, 39–40, 45, 174, 203; FAC, Papers, 19, 28, 30 Dec. 1840, 4, 7, 8–9, 18–24, 26 Jan., 2, 7–8, 12, 14–16, 17–19 Feb. 1841, Papers, Letters to Company directors, 22 Jan., 22 Sept. 1841, ibid, Woollcombe to FAC, 26 June 1840: All FAC items in Jnl APL & Papers TM; Chilman, Diary, 11 Dec. 1841; Creed, 17 Feb., 11 Oct. 1841; *DNZB*, v. 1, 72, 313; Jackson, 5–6, 11–12, 16, 22, 43–47; Jervis, *NZH*, Suppl. 23 April 1892, 1, col. 3; Morley, 86; Mullon, 6; NZC, 3/21, 14 Feb. 1841, 1–9; *NZG & WS*, 26 Dec. 1840, 3, col. 2, 4 Jan. 1841, 6 Feb., 3, col. 1–2, 20 Feb. 1841, 2, col. 3, 20 March 1841, 2, col. 2, 6, col. 4; Quin, 12, 43; R & S, xv, xvi, 64; Scanlan 1985, 20; Seffern, 36–41, 44; Skinner, pamphlet, 2; EJW, 302, 310, 315; WW, Jnl, 12 Dec. 1840–48, 8 Feb. 1841; Ward 1928, 76–77; Wells, 53; Wigglesworth, 91.

1. Carrington was to provide smaller sections than the New Zealand Company's: 2,200 sections at $\frac{1}{4}$ acre each, 1,150 suburban sections of 50 acres each, and 250 allotments. The holders of receipts for land purchase would be allowed to choose by lot, but no one was to be allowed to apply for more than eight allotments, each allotment containing one town section and one rural section. Dr Evans may have provoked this proviso; he applied for forty-five Wellington sections; but at least he was not an absentee owner.
2. Carrington Papers, 28 Dec. 1840.
3. Aubrey's Journals, 2 Feb. 1841, as seen in Wells, 53.
4. It had probably been built for the Rev. Samuel Ironside, who had been in the district in June 1840, escorting home to Ngamotu one group of slaves released by Waikato. He is said to have requested that one house, at least, be built for mission use (Skinner, pamphlet, 2).
5. The surveyors who left Wellington for Taranaki in Aug.–Sept. 1840, after the lottery for the town sections, reported on their return that the inhabitants of the Sugar Loaf Islands were 'interested in whites arriving', and had already begun building one of the large houses. Wakefield had been told by his directors to co-operate with the Plymouth Company, after it was first mooted at the beginning of 1840, and Barrett may have surmised where Wakefield would recommend the West Country settlement be sited. He had had three weeks, since Carrington invited him to join his expedition, to have the other two large whare completed; but he took a colossal gamble on his ability to manipulate Carrington into choosing Taranaki.
6. FAC, Papers, 11–12 Jan. 1841; Seffern, 39; Wells, 53ff.
7. FAC, Letter, in FAC Jnl APL & FAC Papers TM, 22 Sept. 1841.
8. FAC, Papers, 20 Jan. 1841.
9. RB, Will.
10. *NZG & WS*, 6 Feb. 1841, 3, col. 1.
11. Although they could have been among the people Barrett brought back to the ship the next day from Arapaoa.
12. Seffern, 41.

## Chapter 11

Adams, 256–48; RB, Will; Barrett 1962, 2, 8; *BPP*, v. 2, 61–63, 68–69, 70, 88–89, 4–11 June 1844, 277, 24 March 1841, WW to FAC, Part of Appendix to Report from Select Committee on NZ, F No. 7, 629; Burns 1989, 39–40, 45, 165–68, 170, 174, 203, 205; FAC, Jnl, 25 Feb., 1, 5–7, 9–10, 21, 27–28, 30–31 March 6–7, 9–10, 13, 18 April, 4 May 1841; FAC, Letters, in FAC Jnl APL & FAC Papers TM: to WW 8 March 1841, to Cutfield: 7–18 May 1842, 29 June 1841 private letter never sent, 3 Aug. 1841, to NZC directors 22 Sept. 1841, 11, to Thos. Woollcombe Esq. 4 May, 14 Oct. 1841, private letter, 20; Chilman, Diary, 11, 13, 14, 15, 40; Creed, 17 Feb., 11 Oct. 1841, 13 Jan.

275

1843; *DNZB*, v.1, 72, 313; Dieffenbach 1843, 129; Laurenson, 67; LFS, 17, 20, 31; McLean, Papers, 19 Dec. 1846, Aubrey to McLean, v. 5, 9, 14 July 1847, McLean to Sinclair; McLean 1994, 140–41, 149, 156; Moore, 94–114; Morley, 75; Mullon, 12, 18; NZC, Store Account Book; NZC 3/2 1, 1–9, 14 Feb. 1840; *NZG & WS*, 26 Dec. 1840, 3 col. 2, 4 Jan., 6 Feb, 3 cols 1 & 2, 20 Feb., 2 col. 3, 3 April, 3 col. 2, 24 April, 24 July, 3 col. 1, all 1841; Plymouth Company, Confidential Correspondence, 26 June 1840, Woollcombe to FAC, 8 Oct. 1840, Woollcombe to WW, 4 May and 19 June 1841, Cutfield to WW; Quin, 43, 45; R & S, xiv, xvi, 40, 49–53, 63–65, 75, 77, 100; Scanlan 1985, 20; Seffern, 36–44, 47–49, 50, 52–54, 56–58, 60–63, 66–67; Simmonds, 336–42; Tonk in H & N, 36ff, 153; Tullett, 9-10; EJW, 107–108, 302, 303, 310, 315, 340, 364; WW, Jnl, 634; Ward 1928, 76–77, 82, 468; Wells, 53, 57–58, 88; Wigglesworth, 91.

1. *NZH*, 5 March 1892, Suppl. 1.
2. The Barretts' Ngamotu dwelling was valued at only £20 in a census taken five years later.
3. RB later gave evidence that Ati Awa sent £10–12's worth of goods to absent relations in Port IHN, after signing the Ngamotu Deed on 15 Feb. 1840 (OLC, Box 910–914, Case 374D, 22, RB to Clarke jnr).
4. FAC, Papers, 8 March 1841, FAC to WW; *BPP*, v. 2, 70, 88–89, 6 June 1844.
5. FAC, Papers, 9–10 March 1841; *BPP*, v. 2, 63, 6 June 1844.
6. Seffern, 50.
7. Caroline married James Charles Honeyfield in 1852.
8. Barrett always had cannon for his use and they were an integral part of any harbour. One gun fired from an approaching ship produced a guiding pilot from Te Awaiti; Barrett and Love's cannon defended at Otaka pa. Whether Barrett included cannon in his deck cargo on the *Brougham*, or 'bought' one of the *Brougham*'s guns when he arrived is unknown.
9. At a rough estimate, the three houses thus vacated provided about 5,148 sq ft (684 sq m) of living space. With 148 *William Bryan* passengers flooding into the huts, and the Carrington family still living there too, the space per person was less than 6 sq ft (just over ½ sq m); but it was not as tight a squeeze as that because some people were in tents.
10. Whenever vessels called in, journals and logs note that Barrett went out, piloted in, or anchored the craft.
11. Confidential Correspondence of the Plymouth Company, Cutfield to WW, 4 May 1841.
12. The site of the pa Pukeariki, soon to be demolished by the incoming citizens. It is now the area of St Aubyn Street, between Queen and Dawson streets.
13. *NZG & WS*, 20 March 1841, cited in LFS, 16.
14. FAC, Papers, 6–7 April 1841.
15. He really had grossly overdone his claim, however. The area he coveted included land previously 'owned' and cultivated by hapu other than Rawinia's, who had accompanied him and Rawinia on the heke. An unknown factor, or not taken into consideration here, is that Barrett may have had the permission of the local iwi to farm this land, under tuku whenua.
16. It was at about this time that the words 'Maori', or 'normal' and the opposite, 'Pakeha', or 'strangers', began to be used. The immigrants had heard themselves being referred to as Pakeha, and learnt the meaning of Maori.
17. FAC, Jnl, 13 April 1841.
18. Seffern, 50–54.
19. LFS, 20.
20. Chilman, Diary, 11.

## Chapter 12

Barnicoat, Jnl, 8 Feb. 1841; *BPP*, v. 2, 7, 65, 70–71, 75, FAC to Lord Howick 19 July 1844, 677–78 Halswell to WW 4 June 1842; Browse; Burns 1989, 155, 174, 176–77; Carmen, 19; FAC, Jnl, 12 Sept.–25 Dec. 1841; FAC, Letters, (in FAC Jnl APL & FAC Papers TM) 29 June 1841, 3, to Cutfield (never sent), 2 & 22 Sept. to NZC Secretary, 2 & 14 Oct. 1841 to Cutfield; Chilman, Diary, 9–14 Aug., 4, 24 Sept., 2, 9 Oct., 11 Dec. 1841; NZC 108/1 25 May 1840 & 6 Sept. 1841, Hobson to WW, 15 Dec. 1841, Hobson to Stanley; Creed, 17 Feb. 1841; Dieffenbach 1841, v. 1, 1–8; Dieffenbach 1843, 67; Evison, 159–60,

166; Flight, April–July 1842; Heaphy, 70–72; Jackson, 48; Johnson, 43–45; Lawn, 58; Newland, end Dec. 1841; NZC, 5th Report, 31 May 1842, 14; NZC 108 2/52; NZC, 3/1, 8 Feb. 1842, WW to Secretaries; NZC 3/2, Encl. Box nos 248–49, 24 (or 29) Nov. 1841, Halswell to Shortland, 10 (or 16), 24 Dec. 1841, 19 April 1842, Shortland to Halswell, WW to Directors, 18 March 1842; NZC 108 2/19, Scott to WW; NZC 108/1, 25 May 1840, Hobson to WW, 15 Dec. 1841, Hobson to Stanley; NZC, 108 2/59, 4 Sept. 1841, Hanson to Halswell; *NZG & WS*, 24 Aug., 20 Oct. 1841; Plymouth Company, Confidential Correspondence, 8 Nov. 1841 Cutfield to Liardet, 4 Jan. 1842 Cutfield to WW; Quin, 16ff; R & S, 54–56, 58–59, 101–105, 157, 159, 239–41; Seffern, 68, 71–73, 75, 77, 83; Taranaki Report, 23 2.3.2, 26 2.3.4; Taylor, 42; Turton, No. 7, Halswell to Shortland 19 April 1842; Waitt & Bethune; EJW, 300, 377–80, 382–83, 398, 400, 402–403, 459; Ward 1928, 102, 104–105; W. Williams, 8 Dec. 1841; Wood 1843, 40–49.

1. As mentioned previously, the hotel had cost Barrett much more than he expected, but this next unknown debt was the last straw. The refit of the Harriett seems one likely cause. The effect was that Barrett's Hotel remained a well-known Wellington name for the next 150 years, but Barrett had been its owner for less than a year.
2. NZC 108 2/52.
3. FAC, letter to NZC directors (in FAC Papers TM), 22 Sept. 1841, 9.
4. Chilman, Diary, 24.
5. FAC, Papers, 1841–43, 22 Sept. 1841.
6. R & S, 102, 159; Chilman, 23–24.
7. Seffern, 73; R & S, 237.
8. R & S, 55, 103–105.
9. FAC, Papers, 2 Oct. 1841.
10. Barrett had sold his house to be the town of Wellington's library. It is not clear what happened to this arrangement, but a more substantial building was later erected and he obviously hoped to keep the house.
11. NZC 3/2, 24 Nov. 1841, 165–66, Halswell to Shortland.
12. NZC 3/2, 10 or 16 Dec. 1841, 169–71, Shortland to Halswell.
13. Seffern, 72.
14. Also spelt Tupanangi (Carrington, Minutes of Evidence, 70–71, in McLean 1994, 142.
15. R & S, 59.

CHAPTER 13

RB, Will, Codicil, 24 April 1842; *BPP*, v. 2, 496, FAC to Lord Howick, 869, 11 June 1844, FAC evidence; *BPP*, v. 2, Appendix 310, to Report of Directors of NZC, letter no 8 in 43/12, 15 Feb. 1843; Bremner, in H & N, 154; Browse, 20–21 March 1842; Burns 1989, 166, 206, 212–214, 233, 239; FAC, Jnl, 20–21, 27 March 1842, 2, 28 April, 15 June, 2, 7, 8, 23 July, 12, 16, 17, 19, 27 Aug. 1843; FAC, Papers, 9 March 1842, FAC to Woollcombe; Clarke jnr, 1903, 44, 47–49, 51, 54; Clarke, Letters, jnr to snr, 6 Oct. 1841 (1842?); Chilman, Diary, 11 Dec. 1841; *DNZB* v. 1, 131, 314, 402, 490, 522; Dillon, 20–21; FitzRoy 1846, Chapter 3; Flight, 20 Jan., 23, 26 Feb., 1 March, 20 April, 2, 10, 24, 26 28, 30 May, 20 June, 9, 11 July, 7 Aug., 1 Nov., 7, 19, 25 Dec. 1942, 3, 17, 20 Jan., 3, 21, 26 March, 1, 24, 26, 28, 30 May, 9, 14, 15, 21 Aug. 1843; Hursthouse, Diary, 91–96, 13 April–1 May 1843; Lawn, 58–59; LFS, 24, 40, Letter to John Hulse from his sons in Taranaki, 11 May 1842; McLean Papers, Aubrey to McLean, 19 Dec. 1846; McLean 1994, 146, 157; Marjoribanks, 60; Meurant, 8–18 Feb. 1843; Newland, 9 May, 8, 13 July, 27 Aug. 1842, 4 March, 27 Aug. 1843; NZC, Reports, 12th Report, 16–17, WW to Company Secretary, 15th Report; NZC 3/2, 4 June 1842, Halswell to WW, 408–409; NZC, 3/1, 30 May 1842, WW to NZC Secretary; NZC, 3/2, 18 March 1842, Encl. Box 248–49, 3 June 1842, WW to Company Secretary; NZC, 105/1 56–62, 25 July 1842, Wicksteed to WW; OLC, 906, 19–20 May, Wi Tako, Te Puni to Spain, June, 17–18, Te Kaeaea to Evans, 18–19, 11 July 1842, Wairarapa to Evans, Sept., Mahau to Evans, 8–16 Feb. 1843, RB evidence; Parsonson 1978, 175–76; Plymouth Company, Confidential Correspondence, 17 May 1842, 29 Jan., 10 April, 2 July, 22 Aug. 1843, Wicksteed to WW; Quin, 44, 48, 59; Revans, 49; R & S, 51, 59–60, 109, 110, 177–80, 182, 202; Seffern, 91, 92, 94–99, 101, 104–105; Taranaki Report, 27 2.3.4; Tonk, in H

& N, 47–58; EJW, 403, 483, 493, 523, 527–28, 579, 581; WW, Jnl, 20, 24 March 1842; Ward 1928, 58, 301–302; Wells, 93.

1. The directors had actually been aware since it was founded, of New Plymouth's plight, and had sent the moorings before the petition could have reached them.
2. *DNZB*, v. 1, 314.
3. Courtroom details in OLC 906, Box, 45, Case 374, 10 May 1842 .
4. NZC 12th Report, WW to Company Secretary, 16–17, -C- No. 49. -E- Nos 118 to 26. -H- Nos 6, 8.
5. Letter to John Hulse from his sons in Taranaki, 11 May 1842, LFS, 40.
6. Revans, as seen in Burns 1989, 206,
7. Burns 1989, 206–207; Seffern, 97.
8. McLean 1994, 157.
9. Or Britannia, as it was then also known, the name having been transferred from Pito-one.
10. Marjoribanks, 60.
11. Ward 1928, 57.
12. Chapman, Letters, ATL, 3 Feb. 1847.
13. larke jnr, Letter to his father, 25 Nov. 1842.
14. Clarke jnr 1903, 47.
15. OLC 906, 8 Feb. 1843, RB to Evans.
16. Spain's Report, *BPP*, v. 5, 1846, 8; OLC 906, 13 Feb. 1843, RB to Clarke.
17. Remember the reaction of Ati Awa people, sitting in front of the *Tory*'s fire on Sunday 22 Sept. 1839, when news came that Te Rauparaha was going to attack Ati Awa at Kapiti (Chapter 4, 94). OLC 906, 8 Feb. 1843, RB to Evans.
18. Ibid, OLC.
19. Clarke jnr, Letters, to his father, 6 Oct. 1841. In the same letter to his father, who was Protector of the Natives, he also mentioned the injustices he saw dealt out against the Maori, and the fraudulent dealings, which were depriving them of their land. He had '. . . a pretty clear idea or outline of the New Zealand Company cases. They are all even darker than the first case!!' And about Halswell, Commissioner of Native Reserves, he wrote: 'I have some of his reports about the natives and they make me stare! . . . I am quite <u>tired</u>, <u>sick</u>, <u>annoyed</u> and <u>disgusted</u> [*sic*] with Halswell.'
20. OLC 906, 8 Feb. 1843, RB to Evans.
21. OLC 906, 10 Feb. 1843, RB to Evans.
22. Ibid, 10, 13 Feb. 1843, RB to Clarke.
23. Clarke jnr 1903, 48–49.
24. OLC 906, 14 Feb. 1843, RB to Clarke.
25. Clarke 1903, 49.
26. OLC 906, 17 Feb. 1843, Spain to Reihana Rewiti.
27. Ibid.
28. Ibid, 17 Feb. 1843, Reihana to Spain.
29. Ibid.
30. Hursthouse, Diary, 93.
31. Burns 1980, 239–43; Evison, 165–66, Dillon, 20–24. Te Rauparaha and Te Rangihaeata and a large band of men, women and children, using no violence, had prevented surveyors from working on disputed land at Wairau, while they waited for Spain to come and adjudicate on its ownership. Arthur Wakefield, William Wakefield's brother, who was New Zealand company agent in Nelson, together with Police Magistrate H. A. Thompson, had unwisely armed and taken to the area a motley group of labourers to arrest the two chiefs. When Te Rauparaha resisted Thompson's handcuffs, Thompson called up his reinforcements and somehow a shot was fired, no one ever knew by which side, and shooting then 'commenced immediately on both sides' (Burns 1980, 242). Wakefield and Thompson were among the those killed.
32. Written as Tupanangi in McLean 1994, 145.
33. *BPP*, v. 2, 496.

## Chapter 14

*AJHR*, copy of despatch from Spain to FitzRoy, 12 June 1844; Boulton, Papers; *BPP*, v. 5, 1846, 50, 52, 60, 63–64, 69; Burns 1989, 233, 236, 239, 272–73, 348ff; FAC, Jnl, 19 Aug. 1843; *DNZB*, v. 1, 73, 131, 255, 314–15, 470, 482; FitzRoy, 1846, 29; Flight, 11, 13 July, 15 Sept., 6, 14 Dec. 1843; Hulse, 27 Aug., 8 Sept. 1846, 23 Feb. 1847; Jackson, 97–101; Johnson, 43; McLean 1994, 147–48, 157, 158; McLean, Papers, v. 1, 1839–44, v. 2, 29 Nov. 1844, McLean to Sinclair, Enclosure in McLean's Correspondence, v. 3, 147, Wicksteed to McLean, 23 May 1846, v. 5, 9, 14 July 1847, McLean to Sinclair,

# Sources & Notes

Taranaki Case 374 DE; Newland, 29 May, 4, 27, 30 June, 13, 24 July 1843, 3, 5, 28, 29, 31 Aug., 9 Sept. 1844, 14, 19 Jan., 12 Feb., 18 June 1846, 23, 25 Feb. 1847; NZC 105, Despatches of Resident Agent in New Plymouth, Wicksteed to WW, 27 Sept. 1843, 23 May, 30 June 1846, 1, 23 Feb. 1847; OLC 910–914, Box 1, Case 374 D & E, 16 & 17 June 1842, Dorset to WW, 31 May 1844 RB to Clarke, RB to WW & Clarke, Letter from Ironside to Spain, 30 Oct. 1844, Enclosure No. 3, Minutes of Court held at New Plymouth, with Commissioner's Judgement. Final Report, Papers of Sir Donald McLean 1839–77, v. 1 1839–44, 75–80; Parsonson 1978, 44, 47–48, 58; Plymouth Company, NZC Confidential Correspondence, Wicksteed to WW after 1 April 1846; Quin, 93; R & S, 222, 226; Seffern, 103, 105, 107–109, 113; *TH*, 29 March 1941, Centennial Number, 8; Taranaki Report, 27 2.3.4; Wells, 11.

1. Confidential Correspondence, Plymouth Company, 2 July 1843, Wicksteed to WW. Sarah and Caroline were two of the few children receiving any formal education in New Plymouth. In 1843, there were 224 males and 210 females under fourteen years in the town, and only two schools: one, a 'private' school, with three male scholars, the other, Mrs Creed's 'private school for lower classes', attended by ten males and fifteen females (R & S, 226).
2. OLC 910–914, Box 1, Case 374 D (Ngamotu Deed), Case 374E (Taranaki Deed), June 1844, WW to Spain.
3. Seffern, 107.
4. In *These Hundred Acres*, by D. Mullon, there is mention of 'the Fox painting of Barrett'. William Fox came to Wellington in Nov. 1842 and became the New Zealand Company agent in Nelson in Sept. 1843, succeeding Arthur Wakefield after his death at Wairau. He was a prolific artist as well as lawyer and politician (*DNZB*, v. 1, 134–35), and could have worked on a picture of Barrett during his frequent travels around the country. The portrait has no provenance, however.
5. This became known as 'the New Zealand central transaction' (Taranaki Report).
6. OLC 910–914, Box 1, Case 374 D & E, 17 June 1842, Dorset to WW.
7. OLC 910–914, Box 1, Case 374 D, 31 May 1844, RB to Clarke jnr.
8. Known as Wideawake by the Pakeha. He had escorted WW on his trip around the Sounds in 1839, and was drowned in the *Jewess* wreck.
9. Some sources say Barrett acted as interpreter, but Forsaith was the official interpreter, taken to the area by Spain.
10. Monday 3 June 1844, OLC 910–914, Box 1, Case 347 D & E, Enclosure No. 3. Minutes of Court held at New Plymouth with Commissioner Spain's Judgement. Final Report.
11. OLC 910–414, Box 1, Case 374 D, 31 May 1844, RB to WW.
12. 6 June 1844, OLC 910, Box 1, Case 374 D & E, Enclosure No. 3, Minutes of Court held at New Plymouth with Commissioner Spain's Judgement. Final Report, Wahao to Clarke jnr.
13. Ibid, RB to WW.
14. Ibid, 8 June 1844.
15. Ibid.
16. Robert FitzRoy had arrived in late 1843 to fill the gap left by Hobson's death and Britons all over New Zealand had great expectations of him because he had been eulogized as having exceptional qualities for his new job.
17. *DNZB*, v. 1, 500.
18. Clarke, Letters, 27 June 1844; Burns 1989, 271–72.
19. OLC 910–914, Box 1, Case 374 D, 17 June 1842, WW to Evans.
20. Hursthouse 1849, 83–85.
21. NZC 105/5, Wicksteed to WW, 30 June 1846.
22. *TH*, 29 March 1941. In another part of the article, the whaling industry is said to have been under Barrett's direction and was developing in importance and becoming a fruitful source of revenue and commerce to the district. From study of RB's record, this is obviously untrue.
23. Despatches of Resident Agent, Wicksteed to WW, 24 Feb. 1844.
24. Rawinia and the Loves were remembered too, with streets in Ngamotu now bearing their names.

# BIBLIOGRAPHY

### Abbreviations

| | |
|---|---|
| AJHR | *Appendices to the Journals of the House of Representatives, New Zealand* |
| AIM | Auckland Institute and Museum |
| APL | Auckland Public Library |
| ATL | Alexander Turnbull Library |
| AU | University of Auckland Library |
| AUP | Auckland University Press |
| EJW | Edward Jerningham Wakefield |
| BPP | *Irish University Press Series of British Parliamentary Papers: Colonies, New Zealand* |
| H & N | Hamer & Nicholls |
| JPS | *Journal of the Polynesian Society* |
| NA | National Archives, Wellington |
| NZC | New Zealand Company files, National Archives |
| NZG & BS | *New Zealand Gazette & Britannia Spectator* |
| NZG & WS | *New Zealand Gazette & Wellington Spectator* |
| NZH | *New Zealand Herald* |
| NZJ | *New Zealand Journal 1850* |
| OHNZ | *The Oxford History of New Zealand* |
| OLC | Old Land Claims papers, National Archives |
| OUP | Oxford University Press |
| RAHSJ | *Royal Australian Historical Society Journal* |
| RB | Richard 'Dicky' Barrett |
| R & S | Rutherford & Skinner |
| TH | *Taranaki Herald* |
| TM | Taranaki Museum |
| VUP | Victoria University Press |
| WW | William Wakefield |

### Books

*AA Road Atlas of New Zealand.* Hamlyn, Auckland, 1975.

Adams, Peter. *Fatal Necessity.* AUP, Auckland, 1977.

Angus, G.F. *Savage Life and Scenes in Australia and New Zealand.* Smith, Elder & Co., London, 1847. A.H. & A.W. Reed reprint, Wellington.

Atkinson, A. & Aveling, M. (eds). *Australians, 1838 Fairfax.* Syme & Weldon Associates, Sydney, 1987.

Bagnall, A.G. & Petersen G.C. *William Colenso, Printer, Missionary, Botanist, Explorer, Politician.* A.H. & A.W. Reed, Wellington, 1948.

Ballara, Angela. 'Whanganui-a-Tara 1800–1914'. In D. Hamer. & R. Nicholls (eds) *The Making of Wellington.* VUP, Wellington, 1990.

Barrett, A. *The Life of the Rev. John H Bumby.* London, Publisher? 1864.

Beaglehole, J.C. *Captain Hobson and the New Zealand Company.* Smith College Studies, vol. XIII, Nos 1-3 Oct. 1927-Ap 1928, Fay, S.B., Faulkner, H.U. (eds), London.

Beattie, Herries. *The First White Boy Born in Otago.* A.H. & A.W. Reed, Wellington, 1939.

Begg, A.C. & Neil. *The World of John Boultbee.* Whitcoulls, Christchurch, 1979.

Belich, James. *The New Zealand Wars.* AUP, Auckland, 1986.

Best, Elsdon. *The Pa Maori.* Wellington, 1975. Reprint.

Biggs, Bruce. *Maori Marriage.* Wellington, 1960.

Bloomfield, P. *Edward Gibbon Wakefield.* Longmans, Green & Co., London, 1961

Bremner, Juile. 'Barrett's Hotel'. In D.

Hamer & R. Nicholls (eds) *The Making of Wellington*. VUP, Wellington, 1990.
Buller, Rev James. *Forty Years in New Zealand*. Hodder & Stoughton, London, 1878
Burns, P. *Te Rauparaha*. A.H. & A.W. Reed, Wellington 1980
— *Fatal Success*. Heinemann Reed, Wellington, 1989.
Butler, F. *Early Days in Taranaki*. Taranaki Herald Co. 1842.
Butterworth, G.V. & S.M. *The Maori Trustee*. Printed for The Maori Trustee by Through Book Consultants, Wellington, May 1991.
Carkeek, W.C. *The Kapiti Coast*. A.H. & A.W. Reed, Wellington, 1966.
Carman, A.H. *The Birth of a City*. Published by A.H. Carman, 1970.
*Centennial of Taranaki, 31st March, 1941*. New Plymouth Public Library.
Cherfas, J. *The Hunting of the Whale*. Bodley Head, London, 1988.
Clarke, George Jnr. *Notes on Early Life in New Zealand*. J. Walch & Sons, Hobart, 1903.
Colenso, William. *On the Maori Races of New Zealand*. Proof copy, for the New Zealand Institute, Commissioners of 1865 Exhibition in Dunedin, APL.
Condliffe, J.B. *New Zealand in the Making*. Allen & Unwin, Wellington, 1959.
Cowan, James. *The New Zealand Wars and the Pioneering Period*, Vol. I. Govt Printer, Wellington, 1922.
Crawford, J.C. *Travels in New Zealand and Australia*. Trubner & Co, London, 1880.
Cumpston, J. *Shipping Arrivals and Departures, Sydney, 1788-1825*, Parts I, II & III. Published by the author, 24 Holms Crescent, Canberra, Australia, 1963.
Day, Kelvin (compiler). *Shore Whaling*, Series No. 1. Porirua Museum, 1986
*The Dictionary of New Zealand Biography*, vol. I. Allen & Unwin/Department of Internal Affairs, Wellington, 1990.
Dieffenbach, Ernst. *New Zealand and its Native Population*. Smith, Elder & Co., London, 1841.
— *Report to the New Zealand Company*. 1844, *BPP* vol. 2, 556, Appendix, 604-619, F. No. 6.
— *Travels in New Zealand*. Murray, London, 1843.
Dillon, C. *The Dillon Letters, 1843-1853*. A.H. & A.W. Reed, Wellington, 1954.
Doak, Wade. *The Burning of the Boyd*. Hodder & Stoughton, Auckland 1984.
Downes, T.W. *Old Whanganui*. W.A. Parkinson & Co. Ltd. Hawera, 1915.
D'Urville, Dumont. *The Voyage of the Astrolabe*, trans. O. Wright. A.H. & A.W. Reed, Wellington, 1955.
Earle, Augustus. *Narrative of a Nine Months Residence in New Zealand*. Longman Rees, London, 1832.
Easdale, N. *Kairuri: the Measurer of the Land*. Highgate/Price Milburn, Petone, 1988.
Ell, Sarah. *There She Blows*. The Bush Press of New Zealand, Auckland 1995.
Elvy, W.J. *Kei Puta Te Wairau*. Whitcombe & Tombs, 1957.
Evison, Harry. *Te Wai Pounamu, The Greenstone Island*. Aoraki Press/Ngai Tahu Trust Board. Christchurch & Wellington, 1993.

Field, A.N. *Nelson Province, 1642-1842.* Nelson, 1942.
Firth, Raymond. *Primitive Economics of the New Zealand Maori.* Govt Printer, Wellington, 1953.
Gibson, William. *New Zealand Rulers and Statesmen 1840-1885.* Sampson Low, Marston, Searle & Rivington, 1886.
Goddard, R.H. *The Life and Times of James Milsom.* Melbourne, 1955.
Grace, John Te H. *Tuwharetoa.* Reed, 1992.
Grady, Don. *Guards of the Sea.* Whitcoulls, Christchurch, 1978.
— *Sealers and Whalers in New Zealand Waters.* Reed Methuen, Auckland, 1986.
Hamer, D. & Nicholls, R. (eds). *The Making of Wellington.* VUP, Wellington, 1990.
Hankin, Cherry A. (ed.). *Life in a Young Colony.* Whitcoulls, Auckland 1981.
Harnsworth, D.R. *The Sydney Traders: Simeon Lord and His Contemporaries, 1788-1821.* Melbourne University Press, 1981.
Harrop, A.J. *The Amazing Career of E.G. Wakefield.* Allen & Unwin, London, 1928.
Hayter, G.C. *Marlborough Sounds, Tasman and Golden Bays.* Pegasus Press, 1962.
Heaphy, Charles. *Residence in New Zealand.* Smith, Elder & Co., London, 1842, and Capper Press, 1972.
Heberley, Heather. *Weather Permitting.* Cape-Catley, Picton, 1996.
Hood, A. *Dicky Barrett: With His Ancient Mariners and Much More Ancient Cannon at the Siege of Moturoa.*

*Taranaki Herald,* New Plymouth, 1892.
Houston, John. *Maori Life in Old Taranaki.* Reed, Wellington, 1965.
Hursthouse, Charles Jnr. *An Account of the Settlement of New Plymouth, London 1849.* Reprint by Capper Press, Christchurch, 1975.
— *New Zealand, the Britain of the South,* vol. I., Edward Stanford, London, 1857.
Ingram, C.W.N. *New Zealand Shipwrecks, 1795-1975.* Reed, Wellington, 1977.
Irish University Press, *British Parliamentary Papers. Colonies: New Zealand.* Dublin, 1968-1971
Johnson, David. *Wellington Harbour.* Wellington Maritime Museum Trust, 1996.
Keith, Michael. *They Came on The Tides.* Porirua County Council, 1990.
King, Michael. *Maori: A Photographic and Social History.* Heinemann, Auckland 1983.
Lambert, G. and R. *Illustrated History of Taranaki.* Danmore, Palmerston North, 1983.
Langdon, Robert (ed.). *Where the Whalers Went,* Index to Pacific ports and islands visited by American whalers in the 19th Century. Pacific Manuscripts Bureau, Research School of Pacific Studies, Australian National University, Canberra, 1984.
Laurenson, G.I. *Te Hahi Weteriana.* Published as vol. 27, numbers 1 and 2, 1972, *Proceedings of The Wesley Historical Society of New Zealand.*
Lethbridge, Christopher. *The Wounded Lion: Octavius Hadfield 1814–1904.* Caxton Press, Christchurch, 1993.

Letters from Settlers. *Latest Information from the Settlement of New Plymouth, on the Coast of Taranaki, New Zealand.* West of England Board of the New Zealand Company, London, 1842.

*Letters from Settlers and Labouring Emigrants in the New Zealand Company's Settlements of Wellington Nelson and New Plymouth from February 1842 to January 1843.* London, Smith, Elder & Co., 1843.

MacDonald, C.A. *Pages from the Past.* Blenheim, 1933.

McGill, David. *Lower Hutt: The First Garden City.* Wellington, 1991

McMorran, Barbara. *Octavius Hadfield.* Published by the author, Wellington, 1969.

McNab, R. *The Old Whaling Days.* Whitcombe & Tombs, Christchurch, 1913.

Maning, F.E./'A Pakeha Maori'. *Old New Zealand: A Tale of the Good Old Times.* Christchurch, Whitcombe & Tombs, 1922.

Manson, C. and C. *Pioneer Parade.* A.H. & A.W. Reed, Wellington, 1966.

Marais, J.D. *The Colonisation of New Zealand.* London, 1927.

Marjoribanks, Alexander. *Travels in New Zealand.* Smith, Elder & Co., London, 1845.

Marshall, W.B. *A Narrative of Two Visits to New Zealand in H.M. Ship* Alligator, *A.D. 1834.* London, 1836.

Mee, Arthur. *Twixt Tyne and Tees.* Barnsley, King's England Press, 1990.

Miller, J. *Early Victorian New Zealand.* London, OUP, 1958.

Mooney, Kay. *The History of the County of Hawke's Bay*, Part I. Hawke's Bay County Council, Napier, 1973.

Morley, Rev. Dr. *History of Methodism in New Zealand.* McKee & Co., Wellington, 1900.

Morton, Harry. *The Whale's Wake.* Dunedin, 1982.

Mulgan, Alan. *City of the Strait.* A.H. & A.W. Reed, Wellington, 1939.

Mullon, Herbert D. *These Hundred Acres.* New Plymouth, 1969.

Natusch, Sheila. *The Cruise of the* Acheron. Whitcoulls, Christchurch, 1974.

Natusch, S. and Swainson, G. *William Swainson.* Published privately, 46 Owhiro Bay Parade, Wellington, 1987.

Nicholson, Ian H. *Shipping Arrivals and Departures, Sydney, 1826 to 1840*, v. 1. 2, Parts I, II & III. Canberra, 1977.

*NZC Reports*, vol. I, Reports 1–12 (&15th?). 1839–44, presented by Joseph Somes, Esq. to the shareholders. Smith, Elder & Co., London.

O'Connor, I. *Edward Gibbon Wakefield.* Selwyn & Blount, London, 1928.

Orange, Claudia. *The Treaty of Waitangi.* Wellington, 1990.

Orsman, H.W. (ed.). *The Dictionary of New Zealand English.* OUP, Auckland, 1997.

*The Oxford History of New Zealand* — See Rice.

Parsons, R. *Australian Shipowners and their Fleets*, Book V. Magill, Sydney, 1979.

—— *Ships of Australia and New Zealand Before 1850*, vols I & II, Magill, Sydney, 1983.

Parsonson, A. *Nga Whenua Tautohetohe O Taranaki: Land and Conflict in Taranaki, 1839-59*, Revision of Report no. 1 to the Waitangi

Tribunal: 'The Purchase of Maori Land in Taranaki, 1839-59', November 1991.

Petherick, E.A. *Bibliography of New Zealand*. fMs 216, Micro. MS 0213, ATL.

Petre, H.W. *An Account of the Settlements of the New Zealand Company*. London, 1842, Reprint by Capper Press, Christchurch, 1971.

Pevsner, Nikolaus. *County Durham*. Penguin, London, 1953.

Phillips, J. A *Man's Country?* Penguin, Auckland, 1984.

Polack, J.S. *New Zealand: Being a Narrative of Travels and Adventures*. Reprint, Capper Press, Christchurch, 1974.

Prickett, N. *Historic Taranaki: An Archaeological Guide*. Govt Print Books, Wellington, 1990.

Rawson, D.H. *The Restless Mountain*. D.H. Rawson, New Plymouth, 1981.

Rice G.W. (ed.). *The Oxford History of New Zealand*, 2nd edn. OUP, Auckland, 1995.

Richards, Rhys & Chisholm, Jocelyn. *Bay of Islands: Shipping Arrivals and Departures, 1803-1840*. Paremata Press, Wellington, 1992.

Rickard, L.S. *The Whaling Trade in Old New Zealand*. Minerva, Auckland, 1965.

Rogers, L.M. *Te Wiremu: A Biography of Henry Williams*. Pegasus Press, Christchurch, 1973.

— (ed.) *The Early Journals of Henry Williams: New Zealand, 1826–1840*. Pegasus Press, Christchurch, 1961.

Rusden, G.W. *History of New Zealand*, 2nd edn. Melville, Muller & Slade, Melbourne, 1895.

Rutherford, J. & Skinner, W.H. *The Establishment of the New Plymouth Settlement in New Zealand 1841-1843*. Thos. Avery & Sons, New Plymouth, 1940.

Ryan, P.M. *The Reed Dictionary of Modern Maori*. Wright & Carman, Wellington, 1997

Salmond, A. *Two Worlds*. Viking, Auckland, 1991

Scanlan, A.B. *Egmont, The Story of a Mountain*. Reed Wellington, 1961.

— Harbour at the Sugar Loaves: Centennial History of the Taranaki Harbour Board. Taranaki Harbour Board, New Plymouth, 1975.

— *Taranaki, People and Places*. Published by the author, Dist. Thos Avery & Sons, New Plymouth, 1985.

— *Port Taranaki*. New Plymouth, 1991.

Schofield, G.H. (ed.). *A Dictionary of New Zealand Biography*, vols I & II. Dept of Internal Affairs, Wellington, 1940.

Seear, A. (ed.). *Making of New Zealand*. Sheffield House, Wellington, 1989.

Seffern, W.H.J. *Chronicles of the Garden of New Zealand*. Taranaki Herald, New Plymouth, 1896.

Sherrin, R.A.A. & Wallace, J.H. *Early History of New Zealand*. H Brett, Auckland, 1890.

Simpson, Tony. *Te Riri Pakeha*. Alister Taylor, Martinborough, 1979.

Sinclair, Keith. *A History of New Zealand*. Pelican, Auckland, 1988.

Skinner, W.H. *Taranaki Eighty Years Ago*. New Plymouth, 1921.

Smith, S. Percy. *History and Traditions of the Maoris of the West Coast, North Island of New Zealand Prior to 1840*. Thomas Avery & Sons, New Plymouth, 1910.

Strachey, Lytton. *Queen Victoria*. Chatto & Windus, London, 1921.
Sturm, Terry (ed.) *The Oxford History of New Zealand Literature*. OUP, 1991.
Taranaki Education Board/Taranaki Museum. *Dicky Barrett: A Chronicle of a 20 Year Residence in New Zealand*. New Plymouth, 1984.
Taylor, Nancy M. (ed.). *The Journal of Ensign Best, 1837–1843*. ATL monograph, Govt Printer, Wellington, 1966.
Thompson, A.S. *The Story of New Zealand: Past and Present — Savage and Civilized*. Christchurch, 1859.
Tod, Frank. *Whaling in Southern Waters*. New Zealand Tablet Printing Co., Dunedin, 1982.
Tonk, Rosemary. 'A Difficult and Complicated Question', in D. Hamer & R. Nicholls (eds) *The Making of Wellington*. VUP, Wellington, 1990.
Travers, W.T.L. *The Stirring Times of Te Rauparaha*. Wilson & Horton Facsimile, Auckland. 1928 (Orig. 1872).
Tullett, J.S. *The Industrial Heart*. Whitcoulls, Christchurch, 1981.
Turton, Rev. Henry. *Plans of Land Purchases and Maori Deeds of Private Land Purchases 1815–1840*. Grey Collection, APL.
Wakefield, E.J. *Adventure in New Zealand*, ed. Sir Robert Stout. Whitcombe & Tombs, Auckland, 1908.
Walker, Ranginui. *Nga Pepa a Ranginui*. Penguin, Auckland, 1996.
Ward, J.M. *James Macarthur: Colonial Conservative 1798–1867*. Sydney University Press, Sydney, 1981.
Ward, Louis. *Early Wellington*. Whitcombe & Tombs, Wellington, 1928.
Wards, Ian. *The Shadow of the Land*. Govt Printer, Wellington, 1968.
Watene/Watson, Alexander Te Rairini. *Nga Mahara O Mere Ruru Te Hikanui*. Owae Marae, Waitara, Te ra o te Aranga, 1993.
Wells, B. *The History of Taranaki*. Capper Press, Christchurch, 1976.
White, J. *Lectures on Maori Customs and Superstitions*. Govt Printer, Wellington, 1861.
Williams, H.W. *Dictionary of the Maori Language*, 6th edn. Govt Printer, Wellington, 1957.
Williment, T.M.I. *John Hobbs 1800–1883: Wesleyan Missionary to the Ngaphui Tribe of Northern New Zealand*. Govt Printer, Wellington, 1985.
Wilson, J.O. *New Zealand Parliamentary Record, 1840–1984*. Govt Printer, Wellington, 1985.
Wood, John (Lieut). *Twelve Months in Wellington Port Nicholson*. Clowes and Sons, Pelham Richardson, London, 1843; also Pamphlet 120, 1843/1845, ATL.
Wood, R.G. *From Plymouth to New Plymouth*. Reed, Wellington, 1959.
Wright, Harrison. *New Zealand 1769–1840*. Cambridge, MA., 1959.

### Articles & Pamphlets
Barrett, T.M. 'David Scott: Early Flax Trader', *New Zealand Founders Society Bulletin* XXIV, 1962, 2, 8.
Best, Elsdon. 'Maori customs pertaining to birth and baptism', *JPS*, vol. 38: 152, Dec. 1929, 241–69.
— 'Maori agriculture: Cultivated food plants of the Maori, and native

methods of agriculture', *JPS*, vol. 39: 153, Dec. 1930, 346–79.

Busby, James. 'Remarks upon a pamphlet entitled "The Taranaki Question", by Sir William Martin.' Pamphlet 415, 1860, ATL.

— 'The First Settlers in New Zealand and their Treatment by the Government.' Pamphlet, 1856, ATL.

Clarke, George Jnr. Remarks Upon a Pamphlet by James Busby, Esq., commenting upon a pamphlet entitled 'The Taranaki Question', by Sir William Martin, late Chief Justice of New Zealand. Pamphlet 446, 1861, ATL.

Clarke, George Snr. Pamphlet in answer to James Busby's on the Taranaki Question and the Treaty of Waitangi, 1861. ATL.

Elvy, W.J., 'Supposed pit dwellings in Queen Charlotte Sound', *JPS*, vol. 35: 140, Dec. 1926, 329–32.

Downes, T.W., *Maori rat-trapping devices*, *JPS*, vol. 35: 139, Sept. 1926, 228–34.

— 'Bird snaring in Whanganui River district', *JPS*, vol. 37: 145, March, 1928, 1–29.

FitzRoy, Robert. 'Biography: Memorandum', vol. so lettered; otherwise no title. Pamphlet 289, 1852, ATL.

— 'Remarks on New Zealand in February, 1846.' Pamphlet 157 FIT, ATL, 1846.

Foster, J.E. 'Richard Jones of Riley, Jones and Walker, 1787–1852.' *RAHSJ*, vol. II, 1925, 304–305.

— 'Commercial life in Australia a century ago.' *RAHSJ*, vol. IV, 134.

Hinds, S. Latest Official Documents Relating to New Zealand. Pamphlet 49, 1838, ATL.

NZC Reports. Latest information from the settlement of New Plymouth. Pamphlet 104, 1842, ATL.

Power, P.J. 'The Church in Taranaki'; a sketch prepared for the jubilee of the Hawera parish, 1875. Reprinted from the *New Zealand Tablet*, 1925.

Simmonds and Ward, *Colonial Magazine*, June 1846 Lot VIII, London, 158–63, ATL.

— April–August 1846, 336–42, ATL.

Skinner, W.H. 'The Gospel Comes to Taranaki'. Pamphlet, Published & printed by *Taranaki Herald*, Methodist Church of New Zealand Archives, Auckland.

### Newspapers

Relevant issues of the *Christchurch Press*, *Early Settlers Journal*, *Evening Post*, *New Zealand Gazette and Britannia Spectator*, *New Zealand Gazette and Wellington Spectator*, *New Zealand Herald*, *New Zealand Journal*, *Sydney Gazette*, *Taranaki Herald*, *Taranaki News* — all identified in individual sources and notes.

### Letters and Documents

Barrett, Richard. Last Will and Testament, NA, SEP 317, sworn 29 June 1847 (filed under Richard Rundle, 10 August 1847); also 6 Feb 1841, 25 April 1842, 15 Sept. 1862, ATL MS-Papers 2939.

— Journal, ATL, MS-Papers 1736, 1183.

— Letter to his brother, 6 Nov. 1841, ATL MS-Papers 1183.

— Letters to William Wakefield, 31 March 1840, NZC 108 2/52; 14 August 1841, NZC 108 2/52.

Bigge, J. T. Report, Appendix, 701–718, BT Box 13, Mitchell Library, Sydney.
— Report, Appendix, T.H. Armstrong to J.T. Bigge, March 28, 1822, 6524–6525, BT Box 13, 27, Mitchell Library, Sydney.
Browne, G.D. Deposition, February 5, 1831. Enclosure in Despatches from Governor of NSW to British Government, Enclosures 1831-32, p. 922. Rare Manuscripts Dept., Mitchell Library, Sydney, 1267-12 (773?).
Chapman, F.R. Lecture. Under the auspices of the Victoria League; 11 Dec. 1922. ATL.
Chapman, H.S. 'A Walk through Taranaki in 1844.' Lecture delivered at New Plymouth.
— Letters, ATL.
Chilman, Richard. Letter to John Wicksteed, included in Duplicate Despatches of Resident Agent to Principal Agent, 1843–46, Number 35, New Plymouth Library.
Clarke, George Jnr. Letters qMS-0469, ATL.
Creed, Charles. Letters. Methodist Archives, Auckland.
Dakin, J.C. 'The Working Class Pioneers.' MS Papers 2294, ATL.
Darling, Governor. Despatches, 1829, A 1204, CY Pos 773, pp. 970, 975, Rare Manuscripts Dept, Mitchell Library, Sydney.
— Despatches, 1828, 1830. 1202, p. 1006; 1206, p. 346. Rare Manuscripts Dept, Mitchell Library, Sydney.
— Despatches, 1830–31, A 1267-12, pp. 1001-1007. CY 773, Enclosure, letter from R. Jones to NSW Col. Sec. 7 & 8 Sept. 1831. Rare Manuscripts Dept, Mitchell Library, Sydney.
— Despatches, 1831, (Transcripts of Missing Despatches, 1823–32), A 1267 (Part 4) pp. 433, 428, CY Pos 902, Rare Manuscripts Dept, Mitchell Library, Sydney.
— Despatch to Lord Howard, 10 Sept. 1831, 3 enclosures, incl. Jno McLaren and Richard Jones. Rare Manuscripts Dept, Mitchell Library, Sydney.
Duplicate Despatches of Resident Agent, New Plymouth, to Principal Agent of New Zealand Company, 1843–46, New Plymouth Library.
Gipps, Governor. Despatches, 1842, 1228, p. 998, Rare Manuscripts Dept, Mitchell Library, Sydney.
Green, George, 'New Zealand Land Claims', (c. 1869) p. 333. ATL.
— 'James Busby and G. Green's New Zealand's Land Claims', 333. ATL.
Greenslade, W. Methodist Archives, Auckland, White calls on Whitely, 14.
Halswell, E.S. NZC 3/2 p. 165, Commissioner of Native Reserves, to Col. Sec. 24/11/1841.
Harris, Aroha. Crown Acquisitions of Confiscated and Maori Land in Taranaki. Waitangi Tribunal/Dept of Justice, 1993.
Hopper, E. Letters to Mrs Stanhope 1840. ATL.
Jones, Elizabeth. 'Richard Jones, Memories of her father'. Aj 10, Rare Manuscripts Dept, Mitchell Library, Sydney.
Lambton Papers, 140/10, 2 drafts and final copy, ATL.
Lawn, C A. 'The Pioneering Land

Surveyors of NZ.' Published by New Zealand Institute of Surveyors, 1977, Typescript, qMS 1138-1138a, ATL.

McComish, Percy James. Compiler of Barrett family tree. MS-Papers-1561, ATL.

McDonnell, Hilda. 'Trans-Tasman Contacts 1817–1827: The Captain Herd Connection', in Australia and New Zealand: Aspects of a Relationship, Wellington, Stout Research Centre, 8th Annual Conference, Victoria University, Sept. 1991.

McLean, Donald. Papers 1839–77, vol. 1, 3, 4, 5, 12, 1839–54. ATL

— Papers. MS 32, Folder 123, Item 5, ATL.

Marsden Papers, vol. 9, pp. 11–14, D 167. Rare Manuscripts Dept, Mitchell Library, Sydney.

*New Zealand Colonist and Port Nicholson Advertiser.* 2 August, 1824–2 August 1843. Shelf 268/7, ATL.

NZC 108 2/19, Scott, D. to WW; letter re constructing a jetty. NA.

NZC 108, RB to WW, 31 Mar. 1840.

NZC 3/2 pp. 169–71, Colonial Sec. to Halswell, 10/12/1841. NA.

NZC 3/2, pp. 408-09, Halswell to Col. Wakefield 4/6/1842. NA.

NZC 3/2 to 75. Duplicate Inwards Despatches from Principal Agent. NA.

NZC 3/1, 2, 3. 17/8/1839–6/12/41, 940 6059–6061 5/2/1841–21/12/42, 940 6062, 2/1/42–23/12/43. NA. NZC New Plymouth Settlement Papers. qMS-1461, ATL.

NZC Prospectus re Plymouth Company. MS069, TM.

NZC Store Account Book. MS069, TM.

Old Land Claims. OLC 906 (Wellington) 1842-43 (Box 45, Case 374 A, B & C), NA.

— OLC 910 - 914 (Taranaki) 1847. (Box 1, Case 374 D & E), NA.

Parsonson, Ann. Revision of Report No. 1, 'The Purchase of Maori Land in Taranaki, 1839-1859', (Dec. 1991) WAI 143 A1 (a), commissioned by the Waitangi Tribunal.

Partridge, W. Letter to H.S. Chapman in *New Zealand Journal*, 1850, 293.

Plymouth Company, Confidential Correspondence, MS069. Taranaki Museum.

'Report on Outrages committed on the Natives at New Zealand.' In C.O. 201/219, Copied for Aborigines Committee, March 1837, 6 Enclosures. Rare Manuscripts Dept, Mitchell Library, Sydney.

Revans, Samuel. Letters to H.S. Chapman. Typescript, ATL.

Richmond, R. To FitzRoy 21/2/1844, 44/782 and memo from William Spain, c 1A 1 46/1566, ATL.

Society Arts, Manufacturers and Commerce, vol. 8, pp. 83–84, Papers on Manufacturers, 1811–25. London, 1825.

Startup, R.M. 'Taranaki Postal History — An Outline 1841–1968'; paper prepared for the 4 November 1968 monthly meeting of the Taranaki Philatelic Society Inc. ATL.

The Taranaki Report: Kaupapa Tuatahi. Waitangi Tribunal/GP Publishing, 1996.

Waitt & Bethune, Trustees of Waters & Smith, to Superintendent, 25/6/1846 c 1A 1 46/1566, 46/1178. Inwards letters to Colonial Secretary 1A, ATL.

Wakefield, William. Letters 1844, 1847.

MS-Papers-5807, ATL.
Ward, J. 'Instructions for the guidance of W. Wakefield, Principal Agent for the New Zealand Company, after the arrival in New Zealand of the ships about to sail for the first colony.' Ward to W. Wakefield, re Instructions. Letter from New Zealand Land Company to Colonel Wakefield, 16 Sept. 1839, No. 8. MS Papers 704, New Zealand Collection Folder 1, ATL.
— Supplementary information relative to New Zealand, comprising despatches and journals of the company's officers and the first report of the directors. Report on the physical condition and natural history of Queen Charlotte's Sound, Cloudy Bay, Tory Channel, Port Nicholson and the surrounding country.
Waters, Thomas, Smith, James, and Suisted, Charles. Deed of lease, 20// 9/1842 with 1A 1 46/1566. Ground rent £275 per annum; £200 to be spent on repairs, ATL.
Whiteley, J W. Letters. Methodist Archives, Auckland.

DIARIES AND JOURNALS

Barnicoat, J. W. 'Journal of a Voyage from Gravesend to Nelson, 1840–42.' Private collection.
— Diary, 1842–1844. Micro MS 256, typescript, ATL.
Barrett, R. Journal 1827–1847. MS 1736, ATL.
Browse, Captain Nicholas. Diary. MS024, TM.
Carrington, F.A. Journal. NZP Room, APL
— Papers, 1840–1858. NZP Room, APL.
— Papers. MS 001/1 and MS 001/2, TM.
Chilman, Richard. Diary. MS037, TM.
Flight, Josiah. Diary. MS002, TM.
Heberley, James. Journal. MS Papers, ATL.
— Reminiscences, 1809–1843. Micro MS 74, ATL .
Hulse, William. Diary. TM.
Hursthouse, John. Diary. 'A Record of the last months in England, of the Voyage to New Zealand and of the first months in New Zealand.' APL. 2 NZL SH1.
*Letters from New Plymouth*. Subscribers' edn. H.D. Mullon, New Plymouth, 1968.
Knocks, John A. Reminiscences. Typescript, qMS 1117, ATL.
Meurant, E. Diary and letter of Edward Meurant, 1st Oct. 1842–25th July 1847. Typescript, APL.
Moore, George. Journal and Correspondence, 1840–1905. ATL.
Newland, J. Journal. (Copied from Journal in possession of the museum) MS016, TM.
Petre, Mary Ann. Diary, 1842–1844. MS copy micro 0502, ATL.
Snow, Hazel. Notes on Barrett's Journal. MS Papers 2175, ATL.
Wakefield, William. Notes on Journal. Typescript, qMS 2102 ATL.
— Index to Journal. qMS 2103 ATL.
Whitely, Rev. John. Journal 1832–1863. qMS 1832–1863, ATL.
— Missionary to New Zealand, London, 1832. Seen in Kinder Library, St. John's College, Auckland; and in Methodist Church of New Zealand Archives (Auckland) W.M.S. Whiteley Correspondence.
Williams, T.M. MS 104, TM.
Williams, W. MS 024, TM.

## Theses

Boulton, Leanne. 'The Alienation and Administration of Native Reserves in New Plymouth', Case Study. Thesis written for Te Ati Awa, 1997. Unpublished.

Caygill, R.D. 'The Te Awaiti Whaling Station', An Essay on Whaling in and Around Cook Strait, New Zealand. M.A. thesis, Canterbury University, 1948.

Jackson, R.A. 'New Plymouth, Its Foundation and Development'. Thesis in part fulfilment of Diploma of Town Planning, Auckland University, 1968.

McLean, Ronald W. 'Dicky Barrett, Trader, Whaler, Interpreter.' Thesis presented in part fulfilment of an M.A. degree in History, University of Auckland, 1994.

Morton, Henry Albert. 'Whaling in New Zealand Waters in the First Half of the Nineteenth Century.' Ph.D. thesis, University of Otago, 1977.

Parsonson, Ann. 'He Whenua te Utu.' Ph.D. thesis, Canterbury University, 1978.

Poff, Basil John. 'William Fox: Early Colonial Years 1842–1845, Thesis for M.A. degree in History, Canterbury University, 1969.

Quin, B.G. 'Bush Frontier, North Taranaki, 1841–1869: A Study in Economic Development'. Unpublished thesis in part fulfilment of an M.A. degree in Geography, Victoria University, 1966.

Wigglesworth, R.P. 'New Zealand Timber & Flax Trade, 1769–1840.' PhD. thesis, Massey University, 1981.

# INDEX

Anglionby 142, 151
Arapaoa (Arapawa) Island 61, *62*, 63, *81*, 129, 178
Ashdown, George 22
Aubrey, Alexander 176, 182

Baines, surveying assistant 176, 177, 182, 186, 192
Barrett Domain 255
Barrett, Edward 149, 179
Barrett, Hara (Sarah) 73, 86, 117, 123, 125, 141, 164, 169, 174, 185, 188, 189, 215, 235, 254, 255
Barrett, Henare (or Hone Henare Tumehou) 149, 150, 178
Barrett, Kararaina (Caroline) 8, 73, 86, 117, 122, 125, 141, 164, 169, 174, 185, 188, 189, 215, 235, 254, *255*
Barrett, Mereana (Mary Ann) 40, 73, 86, 117, 123, 125, 141, 163, 164, 199
Barrett Rawinia/Wakaiwa 28, 29, *29*, 30, 35, 38, 45, 57, 67, 85, 86, 91, 93, 107, 112, 115, 117, 119, 123, 125, 128, 141, 154, 157, 173, 174, 179, 180, 185, 188, 189, 215, 252, 254, 255, 256
Barrett, Richard 'Dicky'/Tiki Parete 14, 17, 18, 19, 21, *23*, 24, 27, 28, 29, 30, 38, 47, 57, 58, 59, 69, 88, 98, 101, 102, 103, 112, 116, 118, 119, 123, 125, 132, 139, 150, 170, 177, 207; Agent of the New Zealand Land Company 119; Agent for the Natives 143, 145, 157, 238, 256; appearance and character 82, 152, 153, 185, 256–57; and Carrington 171, 173-176, 182, 213–233; at New Plymouth 174, 175, 182, 183, 185, 189, 191, 195, 205, 206, 210, 214, 215, 221, 222, 229, 232, 252, 254; buying Taranaki 124, 125, 127–37; buying Wellington with Wakefield 89–99; family homes 67, 69, 86, 126, 155, 176, 185, 235; finances 144, 157, 166, 169, 172, 180, 192, 193, 194, 199, 200, 202, 208–209, 252, 257; first meeting with William Wakefield 15, 79, 80, 83, 85; funeral 254; heke 52, 56; letters 149, 163–64, 165, 200–201, *201;* management style 69; marriage 188; paying for Wellington 100–124; pilot 89, 174, 250; places named after 255; settling in at Wellington 143, 159, 161–63; and Spain Commission 224–25, 227–31, 238–49; 'tough little warrior' 51; trading 22, 24, 27, 31, 32, 35–36, 37, 40, 53, 58, 68, 73, 74; translating 92–93, 94, 95, 96, 97, 98, 101, 106, 107, 108, 250; travel to Australia 76; tribal life 29–31, 33–34; war and fighting 45–51, 54–55; whaling 59, 60, 62, 63–77, 113, 143, 153, 171, 178, 182, 183–84, 185, 187, 195–99, 202, 209, 214, 221, 232, 237, 247, 248, 251–54, 256–57; Will 179–80, 194; working with Wakefield 87–124
Barrett's Hotel, Wellington, 148, 159, 161, 16*1*, 167, *168*, 172, 179, 180, 199, 200, 204; Taranaki 173, 180, 194, 206, 223, 235
Barrett's Inlet 72, *81*
Barrett's Reef (New Plymouth) 183, 255
Barrett's Reef (Wellington) 111, 112
Bell 232
Bermondsey 17
Best, Lieutenant 167
Blind Bay *See* Whakatu
Bosworth, Bos 22, 66, 182, 198
Boulcott's farm *111*, 252
Brewer, William Vetruvius 180, 216
Britannia town *111*, 139, 142, 146, 147, 149, 151, 154
Brookes, John 115, 121, 162
Brown, Richard 196–98, 213–22, 251, 252, 253, 254
Buckle, Captain 50, 73
Bumby, John 77, 88, 101, 123, 140, 149, 188
Bundy, Billy 22, 51, 66, 182, 198
Busby, James 12

Campbell, Robert jnr 60
Cape Egmont 23, 136
Cape Farewell 75, 119
Cape Terawhiti 137
Carrington, Frederic Alonzo 19, 170, 171, 173, *173*, 182, 184, 186, 187, 189, 191, 205, 208, 213, 216, 218–19, 220, 233, 249, 251; and Barrett 172–79, 190, 193, 202, 213, 233–34
Carrington, Margaret 171
Carrington, Octavius 178, 179, 184, 186, 234, 237
Carrington, Wellington 195, 234.
Chaffers, Edmund M., R.N. 79, 88, 110, 112, 120, 122
Chatham Islands *See* Wharekauri
Church Missionary Society (CMS) 74, 95, 101, 148, 156, 218
Clarke, George jnr. 106, 107, 222, 224, 225, 227, 228, 229, 231, 237, 239–43, 246
Clarke, George snr 203
Cloudy Bay *See* Kauraripe

291

Coglan, G.H. 139
Colonial Office/Colonial Secretary 12, 13, 74, 163
Colonial government in New Zealand 145
Cook, Captain James 12, 23
Cook Strait *See* Raukawa Moana
Cooke, J.G. 220, 222, 232
Council of Colonists 144, 146, 150, 165
Crawford, J.C. 139
Creed, Rev. Charles 185, 188, *188*, 195, 211, 213, 223
Creed, Mrs 'E Mata' 185, 189, 211, 235
Cutfield, George 189, 191, *191*, 192, 196-97, 199, 202, 205-08, 213, 233

Dawson, Captain 205
Deed of sale of land — by Taranaki tribe *135*, 136, 137; Ngamotu 32, 134, *135*, 137, 171, 213, 238, 240, 243, 246; Patea to Wanganui 120, 122; Port Nicholson/'Wanga Nui Atera' 104ff; signed in Queen Charlotte Sound's East Bay 118, 246-47; to Wesleyan Mission at Ngamotu 129
Devon Hotel 223
Dieffenbach, Ernst 19, 79, 98, 110, 124, 126, 127, 128, 131, 135, 137
Doddrey (Doddery/Dodrey), George 78, 100, 102, 131, 133, 135, 145
Dorset, Dr John 79, 98, 110, 112, 130, 131, 137, 238
Duke of Wellington 78, 168
Durham 17
D'Urville Island (Rangitoto) *62*, 121, 177, 178

East Bay 116, 117, 134, 178; East Bay Deed 192
Elmslie, Arthur 78, 79, 80, 85, 87, 113
Eruera Te Puke-ki-Mahurangi, Barrett's father-in-law 28, 30, 53, 125, 127, 129, 134, 185, 193, 212
Evans, Dr George 144, 146, 147, 154, 224-27
Evans, Mrs 146, 157, 161

FitzRoy, Robert 18, 135, 246-50, 252, 254-55
flax (muka) 24, 30, 34, 35, 37, 38, 39, 40, 49, 174, 207
Flight, Josiah 254
Forsaith 240
Fry, Elizabeth 10

'Georgie Bolts', see Toms, Joseph
Gipps, Governor Sir George 114, 128, 136, 144, 151, 165
Grand Fête and Public Ball 158
Grey, Governor George 248, 252, 254-55
Guard, John 38, 59, 60, 61, 63, 64, 65, 66, 75, 88

guns 18, 134, 186, 187, 236, 242, 244, 245
Guyton, William 216

Halse (Hulse), John 233
Halswell, Edmund 203, 210, 224, 225
Hami Parai 109
Hana te Wharetiki 28
Hanson, Richard D. 133, 145, 161, 204, 210, 224
Haowhenua 64, 90, 109
Haupokia 128, 129
Hay, Captain 76
Heaphy, Charles 79, 154, 165, 170, 174
Heberley, James 'Worser' 64, 65, 75, 76, 77, 87, 88, 113, 120, 123, 124, 131, 133, 136, 137, 145
heke 51-56, 57, 60, 62; *see also* Tama-te-uaua
Herd, Captain James 12, 139
Herd's Point 140
Heretaunga River (Hutt) 89, 93, *111*, 141, 143, 149, 153, 159, 170, 252
Hine T.B. 253
Hobbs, John 77, 88, 101, 123, 140, 149, 188
Hobson, Captain William 13, 128, 135, 137, 144, 145, 150, 159, 165, 167, 203, 204, 210, 211, 222, 225, 235, 238, 243, 246
Hokianga/Hokihanga 124, 125, 128, 140
Hone Heke 248
Hone Tanerau *See* Love, Daniel
Honeyfield, James Charles 255
Honeyfield, William John 255
Hongihongi stream & lagoon 26, 49, 51, 185, 222
Honiana Te Puni-kokopu, see Te Puni
Horowhenua 39
Huatoki stream 26, 27, 186, 187, 190, 192, 193, 197, 199, 214, 224, 242
Hunt, U. 180, 194, 211
Hunter, Louisa 150
Hursthouse, Charles 232
Hursthouse, John 232
Hyndes, Thomas 24, 34, 35, 40

Iwikau 54,

Jackson, 'Captain' Jimmy 63, 87, 88, 197
Jervis, H.M. 49, 51, 182, 188, 195, 196
Jones, Richard 66, 68, 73, 75, 82

Kaiapoi 37, 150
Kaipara 84, 119, 125, 130, 133
Kaiwharawhara pa 90, 98, 102, *111*, 159, 218, 225
Kakapo Bay 59, 60, 61, 75
Kapara Te Hau 59, *62*

# Index

Kapiti (region)  17, 32, 38, 39, 41, 42, 52, 54, 55, 56, 57, 59, *62*, 64, 78, 92, 94, 115, 116, 117, 120, 140, 182, 242; island  *62*
Kararaina of Ngai Tahu  149, 150
Kauraripe (Cloudy Bay)  58, 59, 60, *62*, 63, 74, 75, *81*, 113, 198
Kawhia  32, 116, 117, 125, 128, 129, 130, 133, 145
Keenan, William  22, 52, 53, 54–55, 56, 63, 66,
King, Captain Henry  205, 208, 209
Kirihipu Kupapa Kaua  28
Kororareka  101, 248
Kumutoto pa  90, 94, 98, 102, *111*. 143, 156, 165, 225, 231
Kuramai Te Ra  28, 30, 74, 125, 128, 185
Kura te Au (Tory Channel)  61, 69, *81*
Kurukanga  120, 122

Land Commissioner  *See* Spain, William
Lee, 'Black'/'Scipio'  22, 124, 132, 182
Liardet, Captain R.N.  209, 218
Love, Hone Tanerau/Daniel  38, 77, 86, 114
Love, John Agar 'Jacky'/Hakirau  12, 17, 21, 24, 27, 28, 30, 37, 50, 57, 58, 59, 76, 123, 194, 197, 198, 222, 256; becomes ill 77, 78, (death) 113; heke member  52, 56 management style 69; memorial 114, *114* 'the better general' 51; trading 31, 40; war and fighting 45–51, 54–55; whaling 59, 60, 62, 63–77, 87
Love, Mereruru  28, 35, 38, 45, 57, 67

McLean, Donald  247, 248, 249, 251, 253
McLean, Ronald  7, 20
Mahau  93, 94
Mahia  39
Mana Island  *62*
Mananui Te Heuheu Tukino II  54, 55, 62
Manawatu  231
Mangaotuku stream  *26*
Marangai  109
Mataora (Sugar Loaf Island)  43
Mathew, Felton  203
Matipu  42
Matiu (Somes Island)  *62*, 89, *105*, 106, *111*, 138
Mein Smith, Captain William  141, 142, 154, 155, 161, 170, 217,
Mereruru Te Hikanui, see Love, Mereruru
Meurant, Edward  128, 129, 174, 225, 28
Mikotahi (Sugar Loaf island)  *26*, 45, 74, 123, 175
Mohi Ngaponga  109
Mokau  37, 50, 51, 52, 116, 124, 128, 137, 184
Mokau River  33, 130, 136
Moki Te Matakatea  38
mokomokai  55

Moore, George  154, 186, 195
Motueka  177
Motumahanga (Sugar Loaf island)  190
Motuotamatea (Sugar Loaf island)  44
Motunui  32, *33*, 41, 45, 51
Moturoa (Sugar Loaf island)  *26*, 74, 126, 128, 129, 132, 175, 176, 190, 198. *See also* Ngamotu
Moturoa, Te Ropiha  99, 103, 139
Mount Eliot  190, 191, 198
Muaupoko iwi  58, *62*

Nairn, Charles  182, 189, 215, 218, 232
Nelson, Nelson Haven  178, 204, 215, 257
Nesbitt, surveying assistant  176, 182
New Plymouth  178, 180, 194, 196, 200, 205, 207, 211, 214, 218, 223, *223*, 224, 232–34, *236*, 245, 252
New Zealand Association, 1837  12, 13
New Zealand Company, 1825  12, 139; 1834  12
*New Zealand Journal*  152
New Zealand Land Company, 1838 (New Zealand Colonisation Assoc./Company)  12, 13, 17–18, 19, 96, 100, 101, 108, 124, 130, 136, 145, 150, 152, 156, 158, 164, 167, 170, 178, 194, 197, 202, 204, 209, 216, 217, 227, 231, 233–35, 239, 244–46, 248–50, 252, 255
New Zealand Land Claims Ordinance  164
*New Zealand Gazette*  145, 146, 158, 159
*New Zealand Gazette & Britannia Spectator*  158, 167
*New Zealand Gazette & Wellington Spectator*  168, 172, 180, 181, *181*, 194, 214, 231, 254
Nga Rauru  54, 62
Ngaiti  82, 85, 92, 94–95, 103, 106, 115
Ngai Tahu  19, 37, 58, 59, 60, 63, 64, 76, 86, 113, 114, 115
Ngamotu/Nga Motu (place, also Moturoa)  8, 21, 24, 26, 31, 33, *33*, 35, 37, 39, 40, 41, 45, 53, 58, 60, 65, 74, 123, 124, 125, 127, 129, 130, 165, 170–72, 174, 176, 177, 182–84, *183*, 187, 192, 211, 213, 223, 232, 252
Ngamotu Deed  *135*
Ngamotu (hapu)  118, 125, 128, 136, 175, 239, 240, 241, 246, 251
Ngamotu, John  241
Ngapuhi  74
Ngatata-i-te-rangi  28, 43, 90, 109,
Ngati Apa  *33, 62*
Ngati Haua  *33*
Ngati Haumia  91
Ngati Ira  17, 27, 57, 58, *62*, 90, 91, 98,
Ngati Kahungunu  39, 58, *62*, 98, 140
Ngati Kuia  62

Ngati Maniapoto 17, 19, 21, 32, *33*, 36, 52, 131
Ngati Maru *33*, 241
Ngati Mutunga *33*, 37, 41, 55, 74, 82, 90, 91,
Ngati Rahiri 30
Ngati Raukawa 39, 54, 92, 98, 114, 226
Ngati Rauru *33*
Ngati Ruanui 33, 42, 56, *62*, 91, 98, 109, 118
Ngati Tama 32, *33*, 37, 41, 50
Ngati Tawhirikura 27, 54, 157, 236
Ngati te Whiti 27, 28, 157, 220, 236
Ngati Toa 17, 19, 32, 36, 37, 39, 54, 56, 57, 58, 59, 64, 75, 76, 103, 113, 115, 116, 117, 202
Ngati Tuwharetoa *33*, 54–55, 56, *62*, 98, 191, 195
Ngauranga pa 90, 91, 96, 98, 102, 110, *111*, 159, 218, *219*, 222, 225
Nohorua 63

'Old Robulla' 75, 115, 121. *See also* Te Rauparaha
Opunake 33, 128
Orongorongo River *111*
Otaikokako beach 26, 34, 41, 50, 125, 129, 185
Otaka pa 21, *26*, 41, 42, 43, *44*, 46, *47*, 51, 126, 129, *175*, 213, 218, 254
Otakau 64, 113, 185

pa sites — Taranaki (between Moturoa and Te Henui stream) *26*; Whanganui-a-Tara *62*, *111*; Taranaki/Kapiti coast *62*; Totaranui *62*
Parangarau (Parangarahau) 107, *111*
Parininihi (White Cliffs) 33, 176, 240, 241, 249
Paritutu (Sugar Loaf island) 25, 45, 125, 174, 249
Partridge, T.M. 139, 153
Patarutu 27, 30, 31, 37, 40, 50 ,
Patea 55, 120, 124, 218
Patea River *33*, *62*
Pattle, Eliza Anne 9,
Pearson, Captain 150, 151, 154
Pickwick Club 158
Pierce, John 139
Pipitea pa 90, 99, 102, 103, *111*, 141, 142, 143, *148*, 149, 154, 155, 156, 159, 161, *169*, 202, 225, 226, 228
Piri Hapimana *See* Te Mana
Pito-one pa and beach 90, 93, 94, 96, 98, 99, 102, 110, *111*, 138, 142, 144, 146, 153, 158, 169, 218, 219, 222
Plymouth Company 170–72, 177, 189, 191, 193, 194, 208, 209, 233, 257
Poharama 27, 31, 53, 74, 129, 251, *251*, 253
Pomare 74, 90, 91, 154
Porirua 40, 58, 91, 140, 202, 252
Port Hardy 177, 178, 184

Port Nicholson (Poneke) 80, 85, 86, 88, 95, 98, 100, 106, *111*, 112, 127, 129, 130, 133, 137, 138, 152, 163, 207; Wellington purchase deed 104–107
Port Underwood 58, *81*, 113, 115
Puakawa 90, 96, 98, 101, 139, 140, 145
Puke *See* Eruera Te Puke-ki-Mahurangi
Pukeariki pa 27, 43, 248
Pukenamu 54, 56
Pukerangiora pa 21, 41, 42, 43, 51, 127, 129, 241, 242
Pukerua 40, 58, *62*
Puketapu hapu 119, 220, 236, 240, 246

Queen Charlotte Sound/Totaranui *62*, *81*, 178, 197, 239

Rangitane iwi *33*, 58, *62*
Rangitikei River *33*, *62*
Rangitoto/D'Urville Island *62*, 121, 177, 178
Raukawa Moana/Te Moana Raukawa 60, 61, *62*, 71, *72*, 57, 60, 61, 71, 74, 75
Raurimu pa 98, *111*
Reihana (Richard Davis) 101, 109, 148, 156, 163, 230, 231
Rekohu/Wharekauri (Chatham Islands) 74, 83, 90
Revans, Samuel 144, 146, 162, 167, 168, 180, 230
Rimurapa (Rimarap) 103, *111*; Wellington purchase deed 104–106
Rogan, surveying assistant 176, 182
Roka Te Puni 28, 56
Rundle, Richard 221, 232, 247

Scott, David 154, 165, 180, 194, 200, 216
sections and native reserves — Wellington 160, 161–62, 170, 210; New Plymouth 186, 210–13, 219
Selwyn, Bishop 247
Seven Stars Inn 223
Sheridan, Daniel 49, 51, 52
Shortland, Willoughby 151, 159, 162, 163, 167, 203, 210, 230, 231, 235–37
Sinclair Head 137
Smith, James 180, 194, 200, 201
Somes Island *see* Matiu
Spain, William 96, 106, 107, 124, 133, 134, 198; Spain Commission 216, 218, 220, 224, 231, 237–41, 244–45, 247, 248, 252, 255
Spencer, Abel 182, 188, 195, 196
Stewart, Captain James 39, 40
Street, Thomas 24, 35, 39
Sugar Loaf Islands/Sugar Loaves 27, 123, 136, 170, 176, 178, 182, 205, 213, 244
Sydney 24, 31, 52, 58, 59, 72, 73, 76, 77, 113

294

# Index

Taieri Maori Reserve 150
Tainui 21, 32, 33, 41, 45, 131, 238
Taitapu/Taitap 119, 121
Tamaiharanui 37, 39, 64,
Tama-te-uaua 33, 53, *62*, 90, 122, 123, 125. *See also* heke
Taniwa/Taniwha *135*, 244
Taranaki (mountain) *33,* 42, 122, 135, 136
Taranaki (region) 17, 90, 113, 118, 124, 127, 129, 130, 137, 140, 165, 177, 178, 179, 203, 232, 234, 244
Taranaki Deed *135*
*Taranaki Herald* 253–54
Taranaki iwi *33*, 38, 74, 90, 91, 98, 128, 133, 135, 136, 202, 240, 249
Tararua 103, 104–106
Taringakuri (Te Kaeaea) 50, 90, 110, 218, 226
Tautara 28, 30, 38, 53, 56
Te Aro pa 90, 91, 92, 96, 98, 102, 108, 109, 110, *111*, 141, 147, 148, 149, 155, 156, 159, 161–63, 202, 218, 225–28, 231–23
Te Ati Awa 17, 18, 19, 21, 22, 23, *26*, 27, *33*, 54, 74, *81*, 83, 86, 98, 103, 110, 113, 114, 116, 117, 118, 123, 125, 126, 127, 131, 133, 134, 135, 159, 172, 176, 186, 190, 191, 195, 202–04, 207, 219–22, 235–38, 240–50, 255–57; customs and lifestyle 31, 34, 36, 56; guns and war 32, 36, 37, 38, 41, 42, 45–51, 52, 54, 55, 76, 109; migration 39, 63; *see also* Tama-te-uaua; whaling 66
Te Awaiti 59, 60, 61, 63, 64, *64*, 65, 69, 72, 74, 75, 76, 77, 78, 79, *79*, *81*, 88, 113, 114, 116, 143, 171, 178, 182, 198, 256
Te Henui stream 25, *26*, 186, 224, 242
Te Heuheu *See* Mananui Te Heuheu Tukino II
Te Hoiere 62
Te Ika a Maui 17, 22, 57, 58, *62*, 92, 116, 118, *135*
Te Kaeaea *See* Taringakuri
Te Mana (Piri Hapimana) 242, 243
Te Manihera Te Toro 90
Te Matangi 90, 93, 97, 226
Te Moana-a-Raukawa/Raukawa Moana 60, 61, *62*, 71, *72*, 58, 60, 61, 71, 74, 75
Te Moewakatara 242
Te Namu 123, 127
Te Ngahuru 109
Te Peehi Kupe 37
Te Popo 54, 55
Te Puke-ki-Mahurangi *See* Eruera Te Puke-ki-Mahurangi
Te Puni 23, 25, *25*, 27, 28, 31, 37, 43, 53, 57, 86, 100, 108, 152, 162, 165, 202, 218, 226, 256; selling Wellington to Wakefield 89–99
Te Rangihaeata 60, 252

Te Rangitake, Wiremu Kingi 246, 247, 251, 253
Te Rauparaha 32, 36, 37, 39, 40, 41, 56, 57, 58, 59, 60, 63, 64, 75, 76, 86, 91, 94, 103, 114, 115, 118, 120, 121, *121*, 122, 149, 150; horticultural skills 58
Te Ropiha Moturoa *See* Moturoa
Te Ruakotare/Ruatau pa 27
Te Ua 63
Te Uruhi 57, 60, *62*
Te Wai Pounamu 17, 36–37, 38, 40, 41, 59, *62*, 64, 73, 74, 76, 86, 116, 118
Te Whanganui 62
Te Whanganui-a-Tara (Port Nicholson/Poneke/Wellington) 17, 18, 33, 38, 52, 74, 77, 82, 83, 85, 88, 90, 91, 93, 96, 109, *111*, 113, 115
Te Whare 112, 123
Te Wharepouri/Warepouri 23, 24, 25, *25*, 27, 31, 37, 39, 43, 45, 49, 53, 57, 86, 90, 96, 105, 109, 110, 111, 140, 152, 159, 162, 165, 202, 219, 222, 226, 256; arranging payment lots 100, 101, 102, 108; selling Wakefield Wellington 89–99,
Te Wherowhero (Te Weoro Weoro) King Potatau 21, 42, 43, 45, 50, 51, 203, 213, 238, 243
Te Whiti-o-Rongomai 28
Te Wiremu *See* Williams, Henry
Terawhiti 62
Thames River (New Zealand) 12, 140
'The Old Sarpint' 40, 115. *See also* Te Rauparaha
Thompson, H.A. 225
Thorndon 141, 143, 147, 149, 153, 155, 159, 154, 169, 202, 222
Tiakiwai pa 90, *111*, 156, 161, 231
Tohukakahi 46, 48
Toko 109
Tokomapuna 183
Toms, Joseph 63, 64, 68, 75, 153, 197
Toponunga (Tupanangi) 213, 234, 238, 239
Tory Channel/Kura te Au *79*, 86, 115
Totaranui (Queen Charlotte Sound) 52, 61, *62*, 63, 76, 137
Treaty of Waitangi 151, 155, 225
Tribes Taranaki *33*; Taranaki–Kapiti coast *62*; Totaranui (Queen Charlotte Sound) *62*; Whanganui-a-Tara (Port Nicholson/Poneke/Wellington) *62*
Tuarau 123, 124, 127, 152, 240
tuku whenua 99, 103
Turakirae/Turakirai *62*, 98, 103, *111*, 104
Turner, Ellen 9
tutu 210

Wade, George 150, 154, 173, 195
Wade, John 150, 154, 165, 173, 202, 216
Wadestown, Wellington 202
Wahao, Jacob 242, 243
Waiaweka 'Wideawake' 195, 240
Waikanae/Waikanai 42, *62*, 74, 114, 116, 119, 120, 195, 213, 232, 239, 242, 246
Waikato (tribe) 17, 19, 21, 32, *33*, 36, 41, 45, 52, 54, 74, 98, 118, 123, 125, 126, 127, 128, 131, 145, 196, 203, 213, 233, 234, 236, 239, 240, 242
Waipero pa 225
Waiongana/Waiongona River *135*, 213, 242
Wairarapa (chief) 90, 98, 103, 148, 229
Wairarapa (district) 39
Wairau 58, 88, 223, 235, 237
Wairau River *62*
Wairaureki 107
Wairoa River 128
Waitangi pa *111*
Waitangi stream 109
Waitara/Waitara River 21, 33, 41, 42, 53, 128, 133, *135*, 176, 183, 184, 186, 192, 213, 219, 220, 232, 234, 236, 237, 239, 241, 243, 246, 249
Waitohi 38, 78, 114
Waitotara River *33*, *62*
Waiwakaiho River 224
Waiwhetu pa 90, 102, *111*
Wakahingo 241
Wakaiwa *See* Barrett, Rawinia
Wakaroa 106
Wakefield, Edward Gibbon 9, 11, *11*, 18, 144, 156, 168; prison sentence 10; theory of colonisation 10–12, 14
Wakefield, Edward Jerningham 10, 19, 79, 80, 82, *82*, 87, 92, 93, 94, 96, 97, 101, 103, 104, 108, 109, 115, 118, 121, 130, 131, 132, 134, 136, 153, 225, 227, 229, 230; East Bay Deed 118; purchase from Te Rauparaha, deed 116; Taranaki Deed 136; Wellington purchase deed 104–106, 227
Wakefield, Emily, née Sidney 10
Wakefield, Nina 10
Wakefield, William, (Principal Agent, the colonel, 'Wideawake') 9, 14, 15, *15*, 18, 19, 88, 130, 133, 141, 142, 171, 172, 179–81, 187, 203; appearance and character 80, 81, 82, 152; buying Wellington 89–99; dealing with Te Rauparaha 115-16; first meeting with Barrett 15, 78, 79, 80, 86, 87; instructions from directors 83–84; land for NZ Company *16*, *117*; marriage and prison sentence 10, 12; paying for Wellington 100–12; settling in at Wellington 143–170, *168*; Spain Commission 216, 218, 225, 227, 229–31, 234, 237, 238, 243, 246; working with Barrett 91–124
walking boundaries 103, 118
Wallis, the Rev Mr 185, 188
Wanganui district 119, 122, 137, 170, 180, 195, 200, 231, 232
Wanganui iwi *33*, 54–55, 56, *62*, 110, 120, 191, 195
Wangatawa River *135*
Waters, Thomas 180, 194
Waters & Smith 194, 199–202, 210, 211
Weekes, Dr Henry 189
Weller Bros 113
Wellington *105*, *168*, *169*, 182, 196, 197, 199, 202, 210, 211, 216, *217*, 222, 252
Wesleyan Church and mission work 18, 101, 128, 130, 148, 174, 185, 186, 188, 244, 248, *248*, 252
Whakaahurangi track *33*
Whakatu (Blind Bay) 121, 177, 204
whaling 65–73, *71*
Whangaingahau pa 56
Whanganui-a-Tara (Port Nicholson/Poneke/ Wellington) 17, 18, 33, 38, 52, *62*, 74, 77, 82, 83, 85, 88, 90, 91, 93, 96, *105*, 109, *111*, 113, 115, *149*, 229; Wellington purchase deed (Wanga Nui Atera) 104–106
Whanganui River *33*, 37, 54, *62*, 130, 136
Wharekauri/Rekohu (Chatham Islands) 74, 83, 90
White, William 125, 128, 130, 131
Whiteley, John 130, 247
Wickstead, Emma 223
Wickstead, John 218, 219, 220, *220*, 221, 232, 236, 237, 247, 252, 254
Williams, Henry, Te Wiremu 74, 88, 123, 151, 155
Williams, 'Captain', Te Awaiti station carpenter 78, 79, 80, 169
Wiremu Kingi see Te Rangitake
Wi Tako Ngatata 27, 28, 43, 90, 94, 103, 112, 142, 148, 162, 217, 229, 230
Wright, James 51, 66, 153, 159, 198